Airplane
Ownership

Airplane Ownership

Ronald J. Wanttaja

TAB Books
Imprint of McGraw-Hill

New York San Francisco Washington, D.C. Auckland Bogotá
Caracas Lisbon London Madrid Mexico City Milan
Montreal New Delhi San Juan Singapore
Sydney Tokyo Toronto

pbk   3 4 5 6 7 8 9 10 11 12 FGR/FGR 9 9 8 7 6
hc   1 2 3 4 5 6 7 8 9 10 FGR/FGR 9 9 8 7 6 5 4

Product or brand names used in this book may be trade names or trademarks. Where we believe
that there may be proprietary claims to such trade names or trademarks, the name has been used
with an initial capital or it has been capitalized in the style used by the name claimant. Regardless
of the capitalization used, all such names have been used in an editorial manner without any
intent to convey endorsement of or other affiliation with the name claimant. Neither the author
nor the publisher intends to express any judgment as to the validity or legal status of any such
proprietary claims.

Library of Congress Cataloging-in-Publication Data

Wanttaja, Ron.
   Airplane ownership / by Ronald J. Wanttaja.
     p.   cm.
   Includes index.
   ISBN 0-07-068157-0   ISBN 0-07-068158-9 (pbk.)
   1. Airplanes, Private—Purchasing. 2. Airplanes, Private-
-Maintenance and repair. I. Title.
TL671.85.W36 1994
629.133'340422—dc20   94-3509
                                  CIP

Acquisitions editor: Jeff Worsinger
Editorial team: Norval Kennedy, Book Editor
                Robert E. Ostrander, Executive Editor
                Elizabeth J. Akers, Indexer
Production team: Katherine G. Brown, Director
                Ollie Harmon, Coding
                Wanda S. Ditch, Desktop Operator
                Lorie L. White, Proofreading   AV1
Book design: Jaclyn J. Boone   0681589

Contents

Acknowledgments

I'D LIKE TO THANK the following people for their help:

The case study participants, for their candor and willingness to assist. Many thanks to Margaret, Mike, Pablo, Claude, Dan, Stan, Don, John Stephens, and John Ammeter.

Pete Bowers, for inspiration and encouragement. And for letting me fly his darn nice airplane.

Stan Brown and *Ramp Rooster*, for providing a ready camera platform for air-to-air work.

The gang on USENET's rec.aviation.misc, for their suggestions and participation in various surveys.

And once again to Lisa, who provides what every writer needs the most: good editing and chocolate-chip cookies.

Introduction

AMONG THE NONFLYING PUBLIC, aircraft ownership is a sign of the well-to-do. It's definitely a deeply inbred belief. I told a 10-year-old neighbor that I'd take him for a ride in my plane. His response: "Are you rich?"

It's one thing when the nonflying public believes that you have to have bottomless pockets to own your own plane; too many pilots have the same impression. Let's not kid ourselves. Planes aren't *cheap*. You'll never own a plane for the same cost as a well-used car. But just because they cost more than a dented Hyundai doesn't mean you can't afford one.

Aircraft ownership, or access equivalent to ownership, such as a partnership, is available to just about anyone who's willing to work at it. Hard to believe? I have a lot of aircraft-owning friends: one just graduated from college, one's a single mother with two kids, and several are retired. None of them are Wall Street financiers, either; they are just average people who happen to own an airplane.

In this book, we'll explore the costs, benefits, and responsibilities of aircraft ownership. We'll take a look at whether aircraft ownership is for you. We'll explore the process by which you can find the best airplane for the money. We'll examine the ways that any aircraft owner can minimize expenses. And we'll look at how to minimize the effect of problems that occur along the way.

1

Why buy?

IT'S GREAT BEING A PILOT, isn't it? You can come up with a thousand-and-one reasons: the freedom, the vast distances that you can quickly travel, the separation from the everyday world by more than just altitude. But too often there's a question that'll bring you back to Earth.

You'll be talking with some new friends, and it'll somehow come out that you're a pilot. You'll extol the joys of flying and maybe even tell a funny story from your student days. Then someone asks: "Do you own a plane?" When you tell them, "No," there's always an odd little pause. "I rent," you say to fill the silence.

You can imagine what they're thinking. You've just explained how flying is no harder than driving a car, and now you're admitting you have to borrow an airplane for every little flight. Imagine having to beg the use of a car just to go to the store for a loaf of bread.

RENTER'S WOES

Be honest: Do you *like* renting? Sure, a few folks enjoy flying a variety of aircraft and prefer not to be tied down. But face it, renting is the pits. You can't fly when you want. How often have you played the reservation game? "OK, is anything available Sunday afternoon instead of Sunday morning?"

Do you have the hankering to fly something other than plain-vanilla 172s and Pipers? There might be some fun airplanes like Aeroncas (Fig. 1-1) or Cubs available for rent in your area, but their rental rates are often disproportionately high.

Have you ever had to cut short an enjoyable afternoon jaunt to get the plane back to the airport in time for the next renter? Or rush to get home before the FBO closes (Fig. 1-2)? Surely you'd rather own your own airplane. Why don't you?

Two reasons come to mind. The first is cost. Everyone knows airplanes are expensive. Segments of the nonflying public think that we are all fat cats. If your neighbor buys a $30,000 sports car, everyone admires him. If you buy a $10,000 airplane, everyone frowns and mutters that you have too much money for your own good. Yet your neighbor might pay more every year on his sports car's insurance premium than some planes cost to operate for a year, including quite a bit of flying.

Fig. 1-1. Classic aircraft like this Aeronca Chief aren't usually available on the rental market.

Fig. 1-2. If you owned a plane, you could keep it out as late as you wanted.

The second reason you don't own an airplane is related: fear of the unknown. You'd like to know what you're getting into before taking the big step. What'll it really cost? How are you going to handle a sudden maintenance bite? What do you do if you have a forced landing in a farmer's field? I can't help with your neighbors, but I can reduce your anxieties about airplane ownership.

That's what this book is about. I'm going to give you the information to decide if ownership is for you. We'll discuss the costs involved and ways to reduce them. You'll see the whole buying process and the pitfalls that might lurk beneath the rosiest deal. We'll look at maintenance procedures that the owner may perform and how to do them safely. We'll grit our teeth and examine possible actions if ownership turns sour. In short, we're going to look at owning an airplane from the real-world perspective.

To help, I've enlisted the aid of airplane owners. Scattered throughout this book, you'll find a number of case studies. Each includes a detailed, frank revelation of the costs and joys of a particular combination of owner and airplane: a recently graduated college student and his rare taildragger; an artist who won a brand-new airplane in a contest; an engineer who bought a basket-case Stinson for $3,500 and restored it from the ground up; a doctor who commutes to work in a Cessna 182; and an electric company employee and his homebuilt speedster.

By the time you finish this book, you should have a good idea as to whether you can handle and afford owning your own plane. Need convincing? Let's first take a look at a case study that proves you *can* fly for pennies.

CASE STUDY:
ABOUT AS CHEAP AS YOU CAN GET

I get a lot of people mad at me. Sometimes when pilots get together, we talk about how much our airplanes rob our pocketbooks. I usually sit back and listen to the sob stories. Then I chime in with my own numbers: The airplane that I fly costs about $1,000 *per year* to own. That's about $85 per month. Some people pay that much for deluxe cable TV. That gets their attention. My operating expenses are low, as well. The direct costs run about $10 per hour, including fuel, oil, and something in the kitty toward engine overhaul.

I'll fess up: I don't actually own the airplane involved. It's operated by a flying club, but I was the sole member of the club for five years. I handled all the expenses on my own; I know darn well what it took to keep the plane flying. And it wasn't much.

The aircraft is a small homebuilt, the Bowers Fly Baby. The Fly Baby was designed by Peter M. Bowers (Fig. 1-3) in 1960. Bowers is an engineer, author, and photographer. The design won the first Experimental Aircraft Association (EAA) design contest in 1962, and Bowers has sold more than 5,000 sets of plans since.

More than 400 examples have been constructed by amateur builders. Fly Babies come up for sale quite often. I've seen nice examples that have sold for $4,500 or less. (Details of building and buying homebuilt aircraft are discussed in chapter 5.)

Our club's airplane is Bowers' prototype (Fig. 1-4). Our EAA Chapter #26, in Seattle restored N500F in 1982, and he allows us to operate it as a club aircraft. Our

Fig. 1-3. Aviation author and historian Peter M. Bowers designed the Fly Baby homebuilt aircraft.

Fly Baby is about as simple as an airplane can get. It has a small Continental engine. It's made of wood and covered with Dacron fabric. It doesn't have an electrical system or related components: no generator, no starter, no battery, no regulator, no solenoid, no radio, no transponder.

The club shares an open hangar with another homebuilt and pays $62 a month. Simple liability insurance costs $210 a year. State annual registration costs another $25. A good friend with an A&P license performs the annual inspection every year, and club members do the actual repairs and maintenance.

In the seven years I've been associated with the club, our major maintenance expenses were:

- Magneto coil: $85
- Tires: $300
- Replacement inner tube: $80
- Oil cap: $20
- Miscellaneous oil seals: $20
- Repairs from wind damage: $150
- Replacement windshield: $20
- Exhaust pipe ($10)

Fig. 1-4. Operation of Fly Baby N500F is inexpensive because it's so simple: no radio, transponder, electrical system, or doors!

Add that up, and you'll see the plane's average maintenance runs only about $100 or so per year. Because the aircraft is operated in the experimental (amateur-built) category, we were able to save money by using parts from nonaircraft sources (the exhaust pipe came from an auto store) or make parts ourselves (the windshield). Figure 1-5 is a summary of our costs.

The airplane runs quite happily on car gas, burning about $6.50 worth per hour. Add 50 cents per hour for oil and $3 per hour for the maintenance kitty to calculate the $10 per hour rate.

Let's assume you've bought—or built!—your own Fly Baby, and your expenses run the same as our club airplane. Insurance, maintenance, registration, and hangar would come to about $1,060 each year. If you fly 50 hours during that year, you're paying about $31 an hour.

In our club's case, two new members have joined in the past several years. I don't have to pay all the expenses anymore, just $30 a month in dues and the per-hour rate. With two other members in the club, my flying now costs about $17/hour.

THE SORDID TRUTH ABOUT OWNERSHIP

One big fact stands out in the Fly Baby case study: For this rock-bottom case, you'll still spend about $31 per hour for your flying time, which is not much less than the rental rate for a Cessna 152. And that's for a slow, single-seat, and drafty (albeit

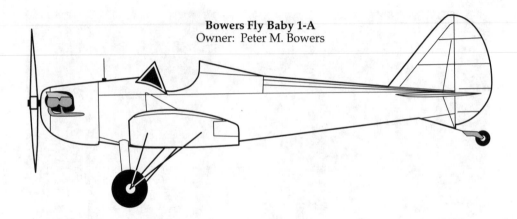

Bowers Fly Baby 1-A
Owner: Peter M. Bowers

Hangar rent: $62/month   Typical yearly maintenance: $100
Insurance (liability): $210/year
Hours/year: 50

Operated by three-person flying club

Fig. 1-5. Summary of Fly Baby ownership costs.

fun) airplane with no radios. I'm going to break the news early: For the average pilot, renting almost always is cheaper than owning.

When I had a Cessna 150, I spent $400 a year for an open tiedown, $500 for insurance, and typically another $500 for the annual inspection and unexpected maintenance. It burned about 6 gallons per hour of $1.80 per gallon 80-octane aviation fuel. That's about $2,000 a year for 50 hours of flying, or $40 per hour. At the time, I could rent a 150 or 152 for $32.

Chapter 2 fully examines costs, but for the moment realize that unless you're flying 100 or more hours per year, renting will be cheaper over the long run. Why buy? Why not just keep renting?

THE KEY TO OWNING

The key to justifying ownership over renting doesn't lie in comparing the per-hour rates. The question is not "Is it cheaper to rent?" The question is "Can I afford to own?" In other words, stop looking at justifying ownership with the prospect of saving money over renting. Instead, decide that you have a new hobby: airplanes.

Compare it to another hobby, perhaps stamp collecting, for instance. How many philatelists' activities are self-supporting? Few, I suspect. Undoubtedly some wheeler-dealer types do turn a profit, but most are content to spend a dollar here and a sawbuck there for a little pleasure. Why should owning a Piper Warrior (Fig. 1-6) be any different?

Fig. 1-6. Consider airplane ownership a hobby rather than a way to escape FBO rental fees.

Unfortunately, unless you buy a 50-cent grab bag at a garage sale and find one of those funny stamps with an airplane printed upside-down, your flying hobby is going to cost a bunch more than stamp collecting. Any hobbyist has to decide how much of their income they can afford to allocate to their favorite pursuit. I don't know how many aircraft are operated by "hobby" owners, but the percentage is high, real high.

It boils down to how much you can afford, or what you're willing to give up. Most airplane owners make some sort of sacrifice. They might forgo buying a new car, or live in an older house or a small apartment, or they vacation in the pilot's seat instead of Europe. Whatever the solution, the money has to be there to keep the plane flying.

Chapter 2 dissects ownership costs to help determine if there is an ownership plan that will fall within your budget. But for right now, you're trying to justify owning. Let's look at some of the advantages.

Ownership advantages

The biggest advantage to owning your own plane is that it's there when you want to fly. Its schedule is entirely subservient to your whims. Say you wake up to a gorgeous Saturday morning, and it is going to be a beautiful day to fly, but you don't own a plane. You dive for the phone and call your local fixed-base operator (FBO)—rats, everything's taken.

If you owned, you'd just drive to the airport because the airplane is ready to preflight and go. And while you're flying, you can go where you wish and return when you feel like it. No one is sitting in an office waiting for you to bring the plane back.

If you want to use an airplane for transportation, owning is even better. Let's say you want to fly to visit a friend 300 miles away, stay two nights, and return. You have to schedule the airplane well in advance, especially if the trip is on a weekend, perhaps two weeks or more at some FBOs.

If the weather is bad the first day, you will have to either shorten the trip or lose your chance. The plane's probably booked on the day following your scheduled return, so you can't just slide the trip back. If you owned, you could leave when you wanted.

To get a solid three-day block of time for a rental airplane, you usually have to pay a certain minimum. It's usually equivalent to around three hours of flight time per day. The roundtrip to see your friend will probably take fewer than six hours flying time, but you would have to pay for nine. The effective per-hour rate just rose by 50 percent.

And if the weather closes in while you've got the plane at your friend's? The FBO keeps charging three hours per day until the plane is returned. And if you have to leave it there to return home? You'll pay for the planes and pilots sent to retrieve the abandoned rental.

Sure, it wouldn't be a picnic if you owned the aircraft. You'd have to arrange to tie the plane down at the destination airport until you could return to collect it. Still, a couple of weeks' tiedown fee is probably $50, or less.

Another advantage of owning has to do with your flying skills and safety. You'll fly more often if you own; a current pilot is a safer pilot. You'll become familiar with the handling and quirks of your own plane. If an emergency situation arises, you'll be more capable of handling it. I had a fuel-contamination engine failure in my old 150 and managed to milk the glide well enough to make a deadstick landing at a nearby airport.

Could I have done as well in a rental airplane? I would like to think so. In the preceding weeks, I had become very interested in glidepath control; I had practiced the techniques in Langewiesche's *Stick and Rudder*. The practice undoubtedly contributed to the successful outcome, whether I'd been flying my own plane or not; however, I believe that my intimate familiarity with the glide characteristics of Cessna Zero-Niner Tango pushed the odds significantly in my favor. All airplanes fly differently, no matter what my fellow engineers say.

Another factor sometimes enters into the formula: mechanical reliability. Some owners are mistrustful of the degree and quality of maintenance that rental aircraft receive and feel that their own planes are more reliable due to their personal attention.

Perhaps. There are FBOs and clubs that scrimp and penny-pinch on maintenance, and there are those that take pride in the reliability of their fleet. Similarly, there are owners who spare no expense when it comes to maintaining their aluminum steed, and there are those who spend just enough to get their plane to pass each annual inspection. Which are you?

In either case, at least your plane won't be subjected to the rough usage that rental aircraft receive. Better reliability results. You'll have more confidence in the aircraft because you'll be familiar with its recent flight and maintenance history.

You'll also be able to optimize the plane to your own needs and desires. Do you require extra cushions to sit comfortably? You can leave 'em in the airplane, rather than dragging them back and forth to the airport. Like to fly from shorter airports? Have the prop repitched to reduce the distance of a takeoff. Want nose art to personalize your machine? Get out the brushes (Figs. 1-7 and 1-8).

Fig. 1-7. You own the airplane. Feel free to add insignia or markings to personalize it.

Fig. 1-8. Heartfelt nose art is worth considering.

There are other, less tangible, benefits to ownership. It's a good excuse to hang around the airport all day. Watching the touch-and-go traffic while sitting under the wing of your own plane is a marvelous way of killing a Saturday afternoon. Ownership is the entrée to a fellowship that includes some of the finest people in the world. A couple of hours spent changing oil and waxing your bird will see a number of owners come by to exchange stories and advice.

Finally, there is often one financial advantage to ownership: Keep the plane maintained and undamaged, and you'll probably sell it for more than you paid. The average general aviation airplane is about 22 years old. Prices for 22-year-old or older airplanes have bottomed out. Some planes with classic appeal even make pretty decent investments.

Ownership disadvantages

If it breaks, it's going to cost big money to fix it. That's probably your biggest worry. It's the fear of the sudden maintenance bill that scares away most potential owners. It can cost $10,000 or more to overhaul some aircraft engines. Scraping together that much money would be tough for most people, especially if they've already gone into hock to buy the plane in the first place.

Breakage is a possibility. We can reduce the chance of major maintenance expenses by a careful prepurchase inspection. We might be able to anticipate and prevent major failures with a thorough maintenance plan. And there are ways to mitigate and stretch out the costs if the worst occurs, which is subsequently explained. Always realize that major failures might occur.

Another disadvantage to ownership is that it essentially restricts you to flying only one airplane. If you own, it's hard to justify the additional expense of renting a plane just for a change of pace. It's especially hard to justify the expense to a spouse.

And, depending upon the aforementioned spouse, it can be a source of family tension. Your mate might not be as fond of flying as you are. An airplane is a hole in the sky into which you throw money; this can produce understandable resentment. Most wives and husbands can come up with excellent alternative uses for the $300 per month that an airplane might cost.

Similarly, if the plane takes up too much of your monthly income, it dulls the joy of ownership. The solution lies in making the most accurate determination of costs prior to purchase and in a realistic assessment of how much you can budget toward ownership.

This book should help. Chapter 2 discusses the financial aspects of ownership and options that can reduce costs. Chapter 3 helps you pick the airplane that fits your needs. Chapters 6 and 7 guide you through the purchase process.

THE FAMILY ELEMENT

Oh, oh. We're going to talk about how your spouse is going to look at your decision to buy an airplane. Maybe you're a carefree single person and don't need to justify your spending. Maybe you make so much money that an airplane won't even dent the family budget. But it probably isn't so.

There's a running joke among the homebuilder clan. Building your own plane requires enormous dedication, often to the exclusion of relationships. Anyway, if you ask a homebuilder how much the plane cost, you'll sometimes get interesting answers like: "$20,000 and my wife" or "$30,000, not including the alimony payments to my ex-husband." You can get a similar response even when you simply buy a plane. It's natural for a wife, husband, or companion to view the "new member of the family" with worry.

My old 150 was a case in point. I bought it just before we moved out of a trailer into a real house. Our new living room stood empty without furniture during the two years that I owned the plane. My wife spoke wistfully of someday filling that empty space. But, bless her heart, she never challenged the aircraft as the obstacle to her dream. The day after I sold the plane, though, she went out and bought $3,000 worth of new furniture.

How can you prevent or minimize friction over your new "toy?" I'm no marriage counselor. But I'm a firm believer in telling the truth up front. Unless the numbers reveal obvious savings, don't justify the airplane by claiming that owning will be cheaper than renting. Even if your numbers show an ownership edge, a mechanical problem can negate all your careful figuring. And if you've bought the airplane "'cause it'll save us money," the hole that you've dug will only get deeper.

Another way of forestalling problems is to buy the aircraft that meets your spouse's level of interest in your new hobby. Say, for instance, you convince him that the plane will be useful for long family trips. If that's the case, *don't buy a small two-seater*. Get something with some room and luggage space. Make very certain that the airplane has room for a child or two, if you have a family.

On the other hand, if your spouse isn't interested in participating, don't go hog wild with the aircraft selection. You won't need a Cessna 206 for zipping out for a hamburger on Saturday afternoon.

Speaking of Saturday afternoons, remember that your spouse might actually be more jealous about your lack of presence than the money spent. Homebuilders know this problem well because they spend hours in the shop, showing up at mealtimes as a bleary-eyed mound of sawdust. Disappearing to the airport every weekend can cause friction. Keep that in mind, lest you end up with no home to go home to.

Tell the truth, and remain attentive and sensitive to your mate's needs, which is a good prescription for any marriage, with or without airplanes.

ANSWERS TO COMMON QUESTIONS

Before we move on, there are several questions many prospective owners have. Let's get them out of the way.

Should I buy a plane to learn to fly in?

If you're a student, or haven't started lessons yet, buying your own plane to take lessons in can seem to be an attractive option because the per-hour rate decreases the more you fly. Availability is never an issue. You'll be able to fly more often and practice whenever you want. However, I don't recommend this course for many reasons.

- Training is hard on an airplane. A bad landing hurts bad enough with the instructor sitting in the right seat of a rental bird. Why put that wear and tear on your own machine? Sure, if you buy a used 150 or 172, it probably has hundreds (if not thousands) of hours in the sweaty grip of student pilots. But why add to it?

- Insurance is going to cost more. Carrying liability would reduce the cost somewhat, but if you carry hull coverage (insurance to cover accident damage) the rates will be higher for a student with 10 hours versus a private pilot with 100 hours.

- It might limit your options for financing. If the plane will be used as collateral on the loan, the bank might be more leery of lending the money to a student pilot.

- Aircraft availability might restrict your training. A renter typically has several airplanes to choose from, and a flight school won't ground the entire fleet for simultaneous routine maintenance. That won't be the case if your airplane requires work. If a part ends up backordered, your lessons might screech to a halt for weeks.

- Picking the right plane is difficult. Did you buy a car before you learned to drive? If you don't have a number of flying hours already, you won't be able to tell bad characteristics from good. Sure, you can ask a flying buddy or your instructor to fly your prospective purchase, but wouldn't you really rather decide for yourself? If nothing else, wait until you've got 30–40 hours under your belt. You'll then have enough background to make an informed choice.

- A good trainer might not be the right airplane for long-term ownership. Cessna 150s are good planes to learn to fly in, but they're rather slow and cramped. If you intend to fly a lot of long cross-countries after you get your license, a Cherokee or 172 would be a better pick.

Note that these drawbacks are mostly intangible; it's more a collection of "What ifs?" than definite iron-clad problems that you'll run into. I've known people who have bought planes to take lessons in, and everything turned out OK.

One case where I do feel that buying a plane for instruction might work out better is if you're going for an advanced rating. If you already have a private and are working on your instrument or commercial, the first three points don't really apply. The last point works in your favor because the larger airplanes generally are better equipped and are better instrument platforms.

Can I save money by buying a plane with a bad engine and installing a converted automobile engine?

This is something I thought about prior to buying my first plane. The engine is the most expensive part of an airplane. If the engine is ruined, the airframe itself can sometimes be picked up at an extremely low price. The homebuilt magazines are full of companies selling Chevrolet, Subaru, Ford, Honda, and other engines converted for aircraft use. Why not put one of these engines in a Cessna or Piper and put it in the experimental category as if it were a homebuilt?

You can't. Oh, there are legal ways of doing this, but none that will work for the average aircraft owner. The experimental category isn't just homebuilts. There are a number of subgroups beneath it, all with different requirements and limitations.

Homebuilts fall into the *amateur-built* group. Amateur-built experimental aircraft have the least amount of limitations placed on their operations. They are subject to tight restrictions during a flight-test period, typically 25–40 hours. After that, they can't be flown over congested areas (except flying to and from airports) and can't be operated for hire (i.e., rented out or otherwise used commercially).

There is one big requirement that must be met before a plane can be certified as experimental/amateur-built: At least half the airplane must be constructed by the builder. Obviously, a Cessna 172 with a Ford engine wouldn't qualify. The FAA allows such an airplane to be placed into the *experimental/research and development* (R&D) or *experimental/market survey* categories.

The R&D category is for testing new aviation concepts, from airfoils to electronics to, yes, engines. Once the modification has proven to be viable, aircraft can move into the market survey category to allow the developers to "test the waters" in regards to placing their new concept in production.

The Cessna shown in Fig. 1-9 is powered by the Stratus EA-81 engine, which is a converted Subaru engine. A Cessna 150 airframe was the most expeditious way to get the powerplant aloft. This 150 is now certified in the experimental/market survey category to allow the owner to take it to airshows across the country to help sell engines to homebuilders.

Fig. 1-9. This Cessna 150 is flying with a Subaru engine. The FAA allows persons or companies developing auto-engine conversions to use certified aircraft as test beds.

If you, like Stratus, *develop* an auto-engine conversion, the FAA is quite willing to give you the regulatory permission necessary to fly the aircraft. If all you want to do is install *someone else's* converted engine, there is no justification for allowing the aircraft into either the R&D or market survey categories.

That's not to say that you can't make *any* changes to your aircraft, but modifications are restricted to those that include approved components. (Chapter 10 reviews the modification process.)

How often does a plane require mechanical inspection?

While renting, you might have noticed that planes occasionally go out of service for two types of inspections: *annual* and *100-hour* inspections. In most cases, a privately-owned aircraft only requires an annual inspection (once a year). The 100-hour inspection only applies to those aircraft operated "for hire." Airplanes provided by your FBO require a 100-hour, but your own plane won't.

Unless, of course, it's on *leaseback* to the FBO. Leaseback is when you lease your plane to a commercial operator for rental or charter. In these cases, the 100-hour is required, and you (as the owner) will have to pay for it.

Paul Poberezny (former president of the Experimental Aircraft Association) has petitioned the FAA for changes to the inspection regulations. He has proposed changing the annual inspection to either biennially (every two years) or every 200 flight hours, whichever is lower. Most lightplane owners will see some cost reduction if his proposal is accepted. Realize that if enacted, it will then become even more important to know your plane and detect problems early.

Should I select a plane based on fuel economy?

Fuel economy is probably more important in car ownership than aircraft ownership. The direct costs of airplane ownership (i.e., fuel and oil) are usually going to be overpowered by the fixed costs (hangar, insurance, maintenance, etc.). While a plane that burns less fuel will naturally cost less to run, the difference might be too trivial to use as a selection criteria. This effect is reviewed in chapter 2.

Can I use the plane in my business?

Of course, as long as your business isn't specifically transporting people or cargo, you can use the plane to support your job however you wish. If a remote site needs a part, you can fly it there. If you've got a business meeting in another town, you can take your own plane.

One of our Fly Baby club's members is an insurance adjuster. He delivers claim checks in our little open-cockpit homebuilt. He doesn't need a commercial pilot license because the aircraft use is incidental to his employment. And in my aviation-writing business, I naturally use the same aircraft in pursuit of stories.

Because they're supporting my business, these flights are tax-deductible. Purchase and operations costs themselves might be deductible in certain circumstances. Consult a tax accountant, though.

The FAA draws the line if you start operating a business that appears to be, or is, an airline or charter service. As long as you don't "Hold out as" (FAA phrase) a transportation service, you should be OK.

CASE STUDY:
AN AERIAL COMMUTER

Few of us like the drive to work. Those of us who are pilots often imagine what it would be like to commute in an aircraft. Let's meet "Doctor Dan," who lives every pilot's dream. When thousands of commuters make their bumper-to-bumper trek to work, Dan wings overhead in his commuter transport, a Cessna 182.

He's the director of a U.S. government research laboratory in Washington, D.C. He started flying in college, paying for his lessons by working in the dining hall and playing guitar in a rock-and-roll band.

He kept flying after graduation from medical school. Eventually, he and his family (his wife is a registered nurse and former TWA flight attendant) moved from San Diego to Washington. Soon after moving, they bought their first airplane, a 1970 Cessna 172. They used it to visit the Atlantic beaches (four hours by car; one hour by Cessna) and other interesting locations on the East Coast. They eventually traded up to a Cessna 182 for the additional speed and range. Dr. Dan sold his 172 for $22,500 and took out a home equity loan to finance the $6,500 difference.

In the meantime, they'd built their dream home on an island in Chesapeake Bay. "The difference between the high-pressure work environment within the Beltway and the rural setting and pace on the island is quite dramatic," he says. "It's like living in Oz and commuting to Kansas." [The Beltway is Interstate 495 encircling the District of Columbia and many of its suburbs. Because many persons consider "inside the Beltway" to be a "power center," high-pressure jobs go with the territory.—Ed.]

Right from the start, Dan planned to commute to work in his Skylane. A bridge connects the island to the mainland, so he didn't have to go by air, but because he already owned an airplane, he decided to give it a try. Two and a half years and 500 flight hours later, he's still commuting in the 182 (Fig. 1-10).

Every weekday he takes off at 5:45 for the 30-nautical-mile trip to an airport near work. Dr. Dan is instrument-rated, but he prefers not to file IFR. Working

Fig. 1-10. If the weather's nice, "Dr. Dan" commutes to work via his Cessna 182.

within the ATC system would add quite a bit of time to his trip. He drives when the weather's too rough to fly, only about 25 percent of the time in the mid-Atlantic states.

One surprising note: His aerial commute doesn't save any time; unless the roads are unusually congested, the trip is 5 minutes faster by car. The 30-mile flight takes only 20 minutes, but extra time is consumed preparing for the flight.

In compensation, though, Dr. Dan escapes the typical pressures and slow-downs of the automotive commute. "You can fly the same old route day in and day out, year in and year out, over 500 times and it's still never the same flight twice," he says. "It never gets old or boring."

Ownership experience

Probably the worst thing you can do to an aircraft or an aircraft engine is to let it sit idle. That's not a problem with Dan's 182. He flies it about 200 hours per year on more than 400 flights, including family vacations. His Skylane is a 1973 Cessna 182P, IFR-equipped including dual navcoms, dual glideslopes, dual marker beacons, two-axis autopilot, and Stormscope.

The aircraft has an STC (FAA approval) to burn automobile fuel. Dan uses it whenever possible. He estimates he's burned more than 10,000 gallons of autogas in his Skylane without any problems. In the last four years, he's had to replace the alternator, voltage regulator, one magneto, muffler, carburetor, ignition harness, two tires, two ignition harnesses, and two batteries.

A lot of maintenance? Not really, for nearly 1,000 flight hours. He performed all the work himself; an A&P provided supervision and signoff. His average maintenance costs run about $750 a year. Active airplanes seem to wear out a lot slower than idle ones rust out. His mechanic is a strong believer in owners knowing their airplanes. He encourages his customers to participate in owner-assisted annual inspections. Working with the A&P/IA, Doctor Dan's annuals average about $350 a year, which is quite a bit less than a typical cost.

He's had—knock on wood—no major avionics problems. One of the display lights on the navcom went out, but the unit was still covered under the manufacturer's warranty. He mailed the unit to the factory and had it back and reinstalled within a week.

He's had one pretty big maintenance expense, though. The Cessna 182 used rubberized fuel bladders from 1956 to 1978. One of Dan's developed a leak. At the time, he was using a canvas cabin cover. The canvas "wicked up" the fuel and held it in contact with the airframe for several days. Severe paint damage resulted. It cost him $2,400 to replace the leaky bladder and repaint the entire left side of the plane.

Another major expense was covered by insurance. One drawback to Dan's commuting is that his plane sits outside unprotected at another field for eight hours every workday. A year ago, someone backed into the tailcone just behind the cabin. The repair cost came to $3,000 and was complicated by slow delivery of replacement structural members from the manufacturer. Fortunately, an IA declared that the aircraft was still airworthy despite the dent, so Dan was able to continue his aerial commute. His insurance company handled the claim without difficulty. His policy's premium runs only about $750 a year, including hull coverage. Because he flies every day and holds an IFR rating, Dan is an insurer's dream.

He rents a hangar on the island's airport for $250 a month. The hangar has electric doors, which is an important factor when you're going to be opening and closing them twice every workday. Dan's expenses are summarized in Fig. 1-11.

Cessna 182
Owner: "Dr. Dan"

Hangar rent: $250/month   Typical annual inspection: $350
Insurance (full): $750/year   Other yearly maintenance: $750
Hours/years: 80

Owner commutes to work in aircraft

Fig. 1-11. A summary of Dr. Dan's operating expenses for his 182.

Advice

"Complexity equals cost," says Dan. He could have bought a small retractable for about the same cost as the 182. He might have gained 20 or 30 knots at cruise. But the timesaving over his 30-mile commute would be minimal. He avoids the higher maintenance and insurance costs of a retractable, plus the 182 has a much heavier useful load, which makes the plane more worthwhile for longer trips.

"Fly it often," he recommends. "Airplanes don't age gracefully if they sit around gathering corrosion." Even though Dan's utilization is higher than average, his maintenance costs aren't out of line. One reason, of course, is his hands-on approach to maintenance. "Knowledge is power. The more you know about the innards of your plane, the more airplane you'll be able to afford," he says.

For those interested in air-commuting: "Chose a commuting arrangement that has a ground-based backup, and don't so-heavily favor the airplane commute that one is tempted to force the issue and go flying when it's dangerous to do so." In other words, if your drive is 2 hours and your flight time is 30 minutes, you're more likely to launch into marginal weather.

Dan also points out another big advantage of commuting by private aircraft. "The greatest boon to pilot proficiency is flying the same plane day in and day out, year in and year out. You learn to sense the airplane's handling and performance." However, "Because a commuter pilot is edging closer to the realm of the do-it-everyday commercial pilot, get an instrument rating. It will do much to improve airmanship, radio competence, and add a safety net for operations that don't go as planned."

2

Money and how to save it

LET'S GET DOWN TO DOLLARS AND CENTS with an in-depth, detailed view of the costs of aircraft ownership, and suggestions to reduce costs. Two types of costs are incurred by the aircraft owner: fixed and direct.

Fixed costs are connected with the ownership, not the operation, of the aircraft. They are predictable, for the most part. Fixed costs are the most irritating aspect of aircraft ownership because they continue whether the plane flies or not. It might snow 30 days out of the month, and on the 31st day the driveway has to be shoveled, but you still have to pay the hangar, insurance, and other fixed costs. For the average owner who flies a hundred hours or so a year, fixed costs are the major expense of ownership.

Fixed costs add up from a number of sources:

- Finance costs
- Insurance
- Hangar/tiedown
- Maintenance
- Miscellaneous (taxes, fees, etc.)

Direct costs change depending upon how much the aircraft is flown. This mainly consists of the fuel and oil expenses.

Maintenance costs are really kind of a hybrid between fixed and direct. Your plane will require an inspection every year, so that's sort of a fixed cost. Yet the cost of the inspection might vary. Similarly, the more you fly the plane, the more things will break down and need repair. Let's take a closer look.

Finance costs

How are you going to pay for your dream machine? If you have to take out a loan to buy your plane (Fig. 2-1), the monthly payments will probably form the largest single cost of ownership. If at all possible, save up for it rather than borrowing.

Fig. 2-1. Few owners are able to pay spot cash for newer aircraft like this Beech F33A Bonanza. Monthly loan payments become a part of the cost of ownership.

A loan will add to your expenses in more ways than just principal and interest. Aircraft loans work similarly to those for cars; the financial institution requires that you carry enough insurance to pay off the loan if the vehicle is damaged. The result is a higher insurance premium.

This won't happen if you finance some other way, such as a signature loan, but the interest rate will be higher. A home equity loan is an attractive possibility, but your spouse might balk at the thought of the family getting evicted if airplane payments are not made. Anyway, you'll still have monthly payments to make. Table 2-1 shows how to calculate the approximate payment based on interest and amount financed.

Don't think you get off the hook if you pay cash, though. An accountant will tell you that paying cash for your airplane costs money, too. They call it "lost interest." If you paid $20,000 cash for the plane, that money could have been invested and earned interest. To an accountant, this represents lost income; hence, it's the same as an expense.

Whether to consider "lost interest" as a real cost of ownership is left to the reader. Some consider it, but many (including myself) don't.

Reducing finance costs

The most obvious way to reduce the finance costs is to win the airplane in a contest, right? Even if we indulge that fantasy for a moment, let's not forget the taxes due after such a windfall. Winning a $100,000 plane would result in approximately a $35,000 tax bill; if you're like me, you would have to take out a loan to pay it. You might end up paying $500 a month on your "free" airplane.

Table 2-1. Loan payment calculation matrix.

Annual Percentage Rate	Loan repayment period and approximate monthly payment per $1,000 borrowed			
	Five Years	Eight Years	Ten Years	Fifteen Years
7.0%	$19.80	$13.63	$11.61	$8.99
7.5%	$20.04	$13.88	$11.87	$9.27
8.0%	$20.28	$14.14	$12.13	$9.56
8.5%	$20.51	$14.39	$12.40	$9.85
9.0%	$20.76	$14.65	$12.67	$10.14
9.5%	$21.00	$14.91	$12.94	$10.44
10.0%	$21.25	$15.17	$13.22	$10.75
10.5%	$21.49	$15.44	$13.49	$11.50
11.0%	$21.74	$15.71	$13.78	$11.37
11.5%	$21.99	$15.98	$14.06	$11.68
12.0%	$22.24	$16.25	$14.34	$11.99

To Use:

1. Take the amount to be borrowed and divide by 1,000
2. Select a repayment period and interest rate
3. Find the intersection of the period and rate, and multiply that amount by the number derived in Step 1

Example: $20,000 borrowed at 10% interest for 8 years:

$$\frac{20,000}{1000} = 20 \qquad 20 \times \$15.17 = \$303.40$$

Loan payment of about $303 per month

On the more realistic side, paying cash is the least expensive solution. Maybe you've got a nice little nest egg tucked away. Maybe a small windfall drops into your lap. A friend of mine sold the house and moved into an apartment to free up the necessary cash. Or perhaps you should just start saving. If you can afford to "own" a basic airplane, you can save up the amount to "buy" the airplane in just a few years.

Need convincing? Let's take my old Cessna 150 as an example. It cost around $2,000 a year to own and fly it. Yet I only paid $6,000 for the aircraft. If I hadn't had the cash to buy it outright, I could have saved up enough in just three years. Sure, this is kind of an unusual case, and you might not want just a little two-seat puddle jumper. But by delaying your decision for a couple of years, you can save a tidy sum toward the purchase of your dream machine. What you have to do is start saving. Now.

Hangar/tiedown

Whether you win, pay cash, or finance your plane, you're going to need somewhere to keep it. Other than loan payments (if any), hangar rent will hog the lion's share of your fixed costs. Tiedowns are cheaper, but hangars provide better protection.

A couple of different types of hangars are available. The most common type is called the T-hangar. The aircraft are aligned in long rows, with the tails of another row tucked between the planes.

A hangar might be nothing more than a long roof with no walls. These protect the planes from sunlight and precipitation, but provide no protection from the wind. Sometimes the individual sections are walled off; add a door and you end up with a closed hangar that protects the airplane from the elements as well as vandalism. Such protection costs more. It might pay off: Our club Fly Baby is kept in an open hangar, and I arrived at the airport one day to find all the propeller nuts missing. Somebody had tried to steal the prop.

There are permutations to the basic hangar pattern. Sometimes, the open T-hangar will have an exterior wall but no internal partitions. You're protected against wind and vandalism (the threat from your fellow owners is small). Or sometimes only partial external walls are erected. The partial walls at least provide a windbreak and a place to set a storage cabinet.

Another type of hangar has a bunch of planes crammed under a big freestanding roof. We'll call these "communal" hangars. They are often heated, which is a heck of a lot better on your aircraft, but you'll pay extra, of course. Plus, your plane usually has to be moved to let some other plane out of the hangar. This leads to "hangar rash," which is minor cosmetic damage occurring when wings bump into your plane. Additionally, you might have to pay a fee to have your plane moved out of and into the hangar.

Whatever type of hangar you find, it's going to cost. The closed hangars at my airport rent for $170 a month—a steal compared to most major metropolitan airports. Hangars 8 miles away at a controlled field run $250 on up. In a year, that's $3,000. If you fly 100 hours a year, the hangar rent alone amounts to $30 an hour.

If your plane can withstand sitting outside, tiedowns are cheaper. The quality varies just like hangars. The lowest-cost is a spot in the grass (Fig. 2-2). You'll probably be expected to install your own tiedown-rope anchors. On the opposite end of the scale will be a marked stall on a concrete apron with solid attach points for your tiedown ropes. In between you will find combinations, for instance an asphalt square parking spot with grass taxiways. This isolates your plane from the moisture, but soggy ground can still affect your operations.

Fig. 2-2. The cheapest tiedown location is on grass, especially at out-of-the-way airports.

Rent varies, depending upon the desirability of the location. Like hangars, tiedowns cost more at close-in major airfields with good security. A tiedown might run $100 a month at these fields, but a grass strip 20 miles away might have spots for $35. I did a survey on Internet regarding tiedown and hangar costs; the results are shown in Fig. 2-3.

Whether an open tiedown works for you depends upon the severity of the local weather and what your airplane is made of. The garden-variety Cessnas and Cherokees are good candidates for outside tiedowns. Fabric-covered planes are less suited. While modern synthetic covering materials last far longer than traditional fabric, they'll last a lot longer under a roof. Older tube-and-fabric aircraft often have spars or other components made of wood. You'd just as soon keep those out of the rain. Grit your teeth and get a hangar.

The cost of renting a hangar might be moot if none are available. The waiting list in my area is 2 years long. At one local airfield, closed hangars almost never become available. If you're seriously considering buying an airplane and you think you'll want a hangar, get your name in now.

Saving money on hangar/tiedown

The best way to save money on tiedown or hangar rent is to shop around. As mentioned above, you can usually find the cheapest prices at the outlying, small fields. But how inconvenient is that to you? When are you going to fly? If you live

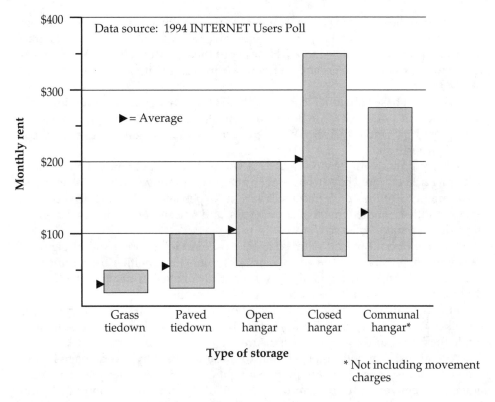

Data source: 1994 INTERNET Users Poll

► = Average

Type of storage

* Not including movement charges

Fig. 2-3. Summarized results of tiedown and hangar cost survey.

right across the street from a major airport, keeping your plane 20 miles away just to save money might be false economy. If you can afford it, bite the bullet and keep the plane as close as possible to home.

The nice thing about hangar/tiedown expenses is that they're something you can check out in advance. A simple phone call to the local fields will reveal the local rental rates and availability. Don't neglect the smaller, private fields in your area. Back when I was learning to fly, our CAP unit kept its Citabria in a barn on a cropduster's private strip. The door opening was actually narrower than the aircraft's wingspan! It took a little tricky maneuvering.

One excellent option for reducing hangar rental is to share the space with another airplane. Hangars are designed for a particular size of aircraft; often two smaller planes can be placed in an area designed for a midsize twin. Hangar rash will be a problem, but because it's just you and one other owner, you'll probably be a bit more careful (further discussed in chapter 9).

Insurance

Aircraft insurance has a lot of similarities to car insurance, but there are some crucial differences that it might pay to be familiar with. Just as with your car, the premium you'll pay to insure your aircraft will depend on a number of factors.

Take experience, for example. Young car drivers with a new license pay higher insurance premiums. Similarly, if you don't have many flying hours, your aircraft insurance premiums will be higher. How about the aircraft itself? A VW bug costs less to insure than a Porsche, and an Ercoupe's insurance bite will be less than that of a Mooney.

Where's it based? If you live in a rough neighborhood, your car premiums are higher. And if you park your Bonanza at an open tiedown on a near-deserted strip, you'll pay more than if it's locked securely in a hangar on a security-patrolled airport.

Finally, how's your driving record? If you've had a lot of fender-benders and speeding tickets, you'll have trouble even getting automotive coverage. And if you do, it might be of the "preferred risk" variety: expensive. Still, most drivers can't escape a little crumpled metal over the years, and a ticket for 37 mph in a 25-mph zone doesn't make you a scofflaw.

But that doesn't hold true in aviation. An FAA violation is a pretty big deal. Accidents are even rarer. The vast majority of pilots have neither bent aluminum nor voided licenses in their pasts. Aeronautical blemishes stand out like sore thumbs and make the pilot subject to higher premiums.

Finally, your rates will depend upon how much insurance you carry. This is just like car insurance; basic *liability* is the cheapest, *comprehensive* covers damage while the car is parked, and *collision* insures against accident damage.

Aircraft liability policies are similar to those of cars. Physical damage is covered by *hull* insurance. Liability policies might cost $250–$500 per year. Hull coverage runs between 3–10 percent of the value of the aircraft per annum. Typical landplane premiums are 5 percent or less; the higher figure is mostly reserved for high-risk aircraft such as floatplanes.

Saving money on insurance

The type of aircraft that you eventually select will have a strong effect on the insurance rates. A four-seat retractable taildragger will have higher premiums than a two-seat fixed-gear trigeared trainer. Of course, the two-seater might not be the right airplane for you. Still, some fixed-gear planes, like the Grumman American Tiger (Fig. 2-4), can nearly match the cruise speed of some light retractables. So don't reject the straightleg planes out of hand.

Your premiums will also be based on a number of other factors, some of which you can't affect. Fundamentally, premiums are set by how many flight hours you have and what ratings you hold. A 50-hour student pilot will pay more than a 1,000-hour private pilot with an instrument rating. If you're ready to buy a plane now, there isn't much you can do.

Some companies reduce your premiums based on recurring training, such as the FAA's "Wings" program. Some companies also take pilot currency into account; if you fly 200 hours per year, an underwriter should consider you a better risk than someone who flies an hour a month. A plane that is flown 200 hours per year is subject to risk more often, but a pilot with plenty of flight hours is more likely to avoid risky problems.

The fundamental solution for lower rates is to shop around. Don't be shy about

Fig. 2-4. This Grumman Tiger is almost as fast as some light retractable-gear aircraft, but its insurance premiums are less.

calling a number of companies. Pick the general type of aircraft that you think you want to buy, and ask for a quote based upon your current pilot qualifications.

Hull coverage has a bit more leeway for minimizing costs. Premiums will be lower if the aircraft is kept in a locked hangar than tied down outside. You can also save dough by selecting a hull policy that doesn't cover the aircraft while it's flying. This *not-in-flight* coverage is designed to give the owner some protection from hazards that have nothing to do with aviation itself. An example would be avionics theft, hail damage, or a hangar fire (Fig. 2-5). If you're willing to bet on your piloting skills, a not-in-flight policy might save you $250–$400 a year; however, if you take out an aircraft loan, the lender will require that you carry full hull coverage.

Chapter 8 expands the discussion about insurance.

MAINTENANCE COSTS

I wish I could tell you what the annual maintenance expense of your plane will be, but my crystal ball has been foggy lately. Some of the costs are "somewhat" predictable. The annual inspection can run from a couple of hundred dollars to several thousand, depending upon the complexity of the aircraft. This is, of course, also affected by how many faults the inspection reveals.

For those that prefer a little predictability to their expenses, *flat-rate* annuals are available. Unless your plane requires extensive, unexpected work, the mechanic performs the inspection and easy repairs for a price that is agreed on in ad-

Fig. 2-5. A leaky fuel system ignited a fire before this airplane's engine was even started. Damages would have been covered under a policy with not-in-flight hull coverage.

vance. The price includes the labor for the inspection and usually a bit of extra labor to fix the garden-variety minor problems. If something big turns up, you'll pay for it. Parts aren't usually included in the price, either.

Flat-rate annuals have supporters and detractors. Some claim the operator will tend to skimp and minimize the labor spent on inspection and repair. Others feel it keeps mechanics from repairing numerous small defects that don't affect airworthiness but can be used to run their profits up. You'll have to make your own decision.

The annual isn't the only maintenance expense. Oil changes will be required at regular intervals (generally 25–50 hours). ELT batteries and certain types of hoses have expiration dates and must be periodically replaced. A transponder must be tested and inspected every two years. The altimeter and static system for IFR aircraft also must be tested every two years.

As far as figuring out the cost of maintenance for your prospective purchase, the complexity of the aircraft must be taken into account. A Bonanza, for instance, is going to cost more to inspect and maintain than a Cherokee 140. Every year, for instance, the inspector will have to place the Beech on jacks and test the retraction system (Fig. 2-6).

Annuals and routine costs aren't the big worries of maintenance. Occasionally things go bad. Very bad. A friend had to replace two engine cylinders in his Tri-Pacer. The total cost came to almost $2,500.

Fig. 2-6. The FAA requires that the gear mechanisms on retractable-gear aircraft be tested at every annual. This increases the cost of the inspection.

It isn't just parts wearing out, either. Manufacturer service bulletins and FAA airworthiness directives (ADs) can add to unanticipated maintenance expenses. Aircraft manufacturers track problems with aircraft. A component might be improved if a certain problem keeps recurring in a particular model. The availability of the component and the underlying problem will be announced as part of a *service bulletin* from the manufacturer.

Compliance with a service bulletin is optional; however, if the problem is especially safety-related, the FAA will issue an *airworthiness directive*. Compliance with an AD is mandatory.

The best way to handle the unexpected maintenance expense is to establish a maintenance reserve by setting aside a fixed amount every flight hour. The basic purpose of this fund is to pay for the eventual overhaul of the engine. It also comes in handy for other expensive repairs.

Determining the rate to "charge" yourself for the maintenance reserve is relatively easy. It's based upon the *time between overhauls* (TBO) of the engine and the estimated cost of overhaul. Typical engine TBOs range from 1,600 to 2,200 hours; complete overhaul will cost several thousand dollars.

To determine the hourly rate for the maintenance reserve, take the overhaul cost and divide by the hours until overhaul (at the time you bought the plane). It's as simple as that. The Lycoming O-235 shown in Fig. 2-7 has a 2,000-hour TBO. Count on an overhaul cost of $5,000, which is rather low, and build your maintenance reserve at the rate of $2.50 a flight hour. If you buy a plane with a Lycoming with 1,000 flight hours, the rate jumps to $5 an hour because the overhaul charge must be "earned" over only 1,000 flight hours.

Fig. 2-7. The Lycoming O-235 engine is used on a variety of aircraft. Textron Lycoming Reciprocating Engine Division, 652 Oliver Street, Williamsport, PA 17701

The maintenance reserve should be deposited into a separate account to keep it isolated from other needs. Of course, it is available for worthier household or emergency needs that might arise. Or, you may decide to ignore the maintenance reserve. Most owners don't keep their aircraft that long; you'll probably sell the plane long before overhaul time. Being more grasshopper than ant, my intent has always been to dig up the money from somewhere when the time comes.

Keep in mind that TBO is a nebulous term; chapter 3 has more information regarding engines. With ordinary luck, there shouldn't be many unexpected maintenance hits on your budget. I asked some friends what their annual maintenance bills typically came to, and the results are in Fig. 2-8. As the saying goes, "your mileage may vary."

Saving money on maintenance

If you've got money to burn, go ahead and have an A&P do all the maintenance on your airplane. But if your tender's too tight for tinder, do as much of the work as you can. The Federal Aviation Regulations state (in part): "The holder of a pilot certificate . . . may perform preventative maintenance on any aircraft owned or operated by that pilot which is not used under Parts 121, 127, 129, or 135." That's from FAR 43.3. The term *preventative maintenance* is defined in appendix A of FAR 43 and summarized in Fig. 2-9.

If you glance at the list, there's a lot of stuff you can legally do! Change the oil. Replace tires and tubes. Grease wheel bearings. Patch fabric and metal. Replace bulbs. Clean and gap spark plugs.

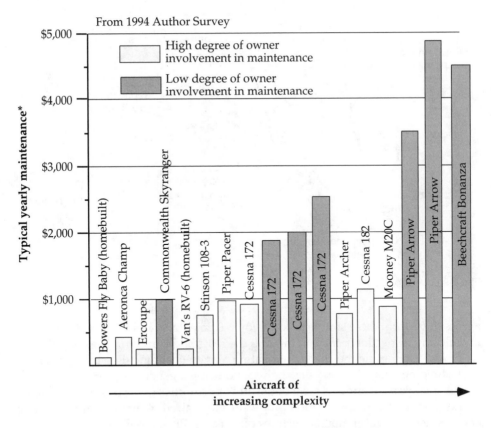

Typical yearly maintenance*

High degree of owner involvement in maintenance

Low degree of owner involvement in maintenance

$5,000

$4,000

$3,000

$2,000

$1,000

Bowers Fly Baby (homebuilt)
Aeronca Champ
Ercoupe
Commonwealth Skyranger
Van's RV-6 (homebuilt)
Stinson 108-3
Piper Pacer
Cessna 172
Cessna 172
Cessna 172
Cessna 172
Piper Archer
Cessna 182
Mooney M20C
Piper Arrow
Piper Arrow
Beechcraft Bonanza

Aircraft of increasing complexity

* Based on owner reports. Data includes annual/100 hour inspections, other periodic procedures, plus airframe and engine accessory repairs as required. Does not include major airframe or engine repairs (overhauls, cylinder replacement, etc.) nor discretionary procedures such as painting or interior refurbishment.

Fig. 2-8. Some typical annual maintenance bills for a variety of owners.

How much can you save? Let's check out an oil change. A small Continental engine's oil must be drained and replaced every 25 flight hours. It'll probably cost you about $50 to have that done by an A&P. The oil itself is $2.50 or so a quart. For a fun afternoon of messing with your plane and getting greasy, you've saved about $40. If you fly 100 hours a year, that's more than $150 you've saved on oil changes alone. That's just one aspect of owner maintenance; chapters 11 and 12 provide the background necessary for performing common preventative maintenance.

Of course, there are a lot of other items that must be left to licensed mechanics. The annual inspection is the big one. One way to save money is with an "owner-assisted" annual. The inspection involves a lot of grunt work that doesn't require a licensed mechanic to perform. Many facilities reduce the price if the owner

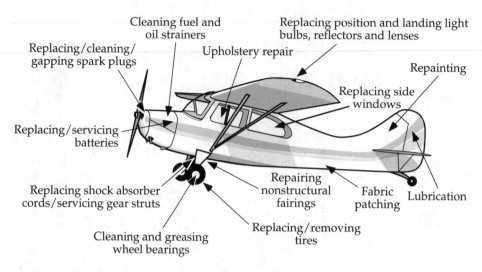

Cleaning fuel and oil strainers

Replacing position and landing light bulbs, reflectors and lenses

Replacing/cleaning/gapping spark plugs

Upholstery repair

Repainting

Replacing side windows

Replacing/servicing batteries

Replacing shock absorber cords/servicing gear struts

Repairing nonstructural fairings

Fabric patching

Lubrication

Cleaning and greasing wheel bearings

Replacing/removing tires

Fig. 2-9. FAA-approved owner maintenance operations.

agrees to perform these simple tasks. If you participate, you'll do jobs like opening and closing inspection panels, removing the seats and carpeting, washing the exterior and vacuuming the inside, lubricating hinges, changing light bulbs, fetching parts, holding propellers and flashlights, and being a general helper.

Owner-assisted annuals are wonderful opportunities to see how your airplane works and to learn the proper procedures for aircraft maintenance. They're cheaper, too! However, not all facilities offer them. Plus, the FBOs generally only operate Monday through Friday, so you'll have to take time off work to help.

Finally, it doesn't hurt to shop around when selecting A&Ps to work on your plane. Often, mechanics in the small, outlying airports have lower rates and are more willing to allow the owner to participate.

DIRECT COSTS

Direct costs are easy to compute. Take the engine's per-hour fuel burn, and multiply by the cost of fuel. Add a buck or so for oil, and there are the direct costs.

But there is one major variable in the direct cost equation: the cost of gas. The solution is simple. Buy an airplane with an engine capable of running on 80 octane avgas. Of course, 80 octane is rare these days, but autogas isn't. Most 80-octane airplanes can be approved to run on ordinary car gasoline. As I write this, unleaded car gas is $1.10 a gallon while 100LL is $1.95. Plus, I can apply for a 10-cent per gallon refund on the state road taxes included at the pump.

There are drawbacks. The logistics can be frightful, depending upon your aircraft. It's not as hard to haul gas for a Luscombe than it is to keep a Cessna 182's tanks brimming with autogas. While some FBOs sell auto fuel at the airport just like avgas, you'll probably have to haul gas from the local service station. Carrying autogas and filling an airplane's tanks are fraught with obvious hazards.

To legally use car gas, the aircraft must have received a supplemental type certificate (STC). The Experimental Aircraft Association and Peterson Aviation have performed tests to the FAA's satisfaction on dozens of aircraft types. The STC is not a blanket approval for all flying airplanes; you must purchase the STC from either EAA or Peterson, and that single STC applies to only your airplane. For most airplanes, the only modifications required are a logbook entry and the replacement of the fuel grade sticker next to the gas cap.

What will happen if you use car gas without the STC? If the FAA catches you, it's a violation. Unfortunately, the way they usually catch violators is in a post-crash fuel analysis. Probably your biggest worry is your insurance company because if you have an accident, and you were using autogas *without the STC*, the insurance company might be able to deny your claim.

The other major direct cost is oil. Engines typically burn a quart every 5–10 hours. The oil should be changed every 25 hours for engines without oil filters, or at 50-hour intervals if an oil filter is installed. Aviation oil costs $2 or so a quart; hence, the oil costs come to about a dollar an hour.

ADDING IT UP

Table 2-2 shows typical costs based on various annual utilization and the autogas/avgas decision. Fixed costs are the major expenses until utilization reaches about a hundred hours per year. One hundred hours per year works out to about two hours a week; about eight hours a month. That's quite a bit of flying, especially for knock-around fun planes like Champs or 150s.

Your actual utilization will probably lie between 25 and 75 hours a year unless there are specific, regular flights you plan on making. The first year will be greater, of course. Do the numbers shock you? Are they completely out of reach? If so, stay tuned because there's a way to cut those numbers by 50 percent or more.

PARTNERSHIPS: SHARING THE FUN

Assume you've worked out that your dream plane will cost about $400 a month to own. Too much? Well, if you bought half the airplane, your expenses would drop drastically, wouldn't they? All you have to do is find someone to buy the other half of the airplane.

It makes a lot of sense. Most of the expense of ownership comes from the fixed costs, not the direct ones. Let's say you fly 100 hours a year, and your fuel cost is $1,000. Assume that the fixed costs for the plane run another $4,000. If you owned the plane yourself, the plane would be costing you $50 an hour.

With a single partner, your part of the annual fixed costs drop to $2,000. You'd still be paying the same direct costs, but your plane just got $20 an hour cheaper to run. With a third partner, that drives your rate down to about $24 an hour, which is less than half the original amount.

On the downside, the airplane's not exclusively yours. That can dull the attractiveness for some, but it's still better than renting. Do you need the plane for a whole week? Just call up your partners and clear it with them. Have to leave it at

Table 2-2. Sample costs of ownership.

Case 1: Simple two-seater, outside tiedown on country airfield, paid cash
Case 2: Basic four-seater, paved tiedown at city airport, paid cash
Case 3: 4-seat retractable, closed hangar, $25,000 ten-year loan @10%

	Yearly				Monthly		
	Case 1	Case 2	Case 3		Case 1	Case 1	Case 3
Storage	$300	$840	$3,000		$25	$70	$250
Insurance							
Liability	$250	$350	$400		$21	$29	$33
Hull	0	$600	$700		0	$50	$58
Loan payments	0	0	$3,960		0	0	$330
Annual	$300	$750	$1,500		$25	$63	$125
Maintenance	$500	$1000	$1,500		$41	$84	$125
Total fixed & maintenance	$1,350	$3,540	$11,060		$113	$295	$921

	Case 1	Case 2	Case 3
25 Hours/year			
Fuel cost	$165	$390	$490
Total cost	$1,515	$3,930	$11,550
Cost per hour	$61	$157	$462
50 Hours/Year			
Fuel cost	$330	$780	$980
Total cost	$1,680	$4,320	$11,680
Cost per hour	$34	$87	$240
100 Hours/year			
Fuel cost	$660	$1,560	$1,960
Total cost	$2,010	$5,100	$12,660
Cost per hour	$20	$51	$130
200 Hours/year			
Fuel cost	$1,320	$3,120	$3,920
Total cost	$2,670	$6,660	$14,620
Cost per hour	$13	$33	$75

Assumptions: Autofuel @$1.10/gal 100LL @ $1.95/gal
Fuel consumption: Case 1: 6 GPH autofuel
Case 2: 8 GPH 100 LL
Case 3: 10 GPH 100 LL

another field due to bad weather? Ditto. And if an unexpected maintenance bite hits, the cost gets spread out across many shoulders instead of just yours.

Partnerships versus clubs

The terms partnership and club are sometimes used interchangeably. The difference is subtle, yet important. In a *partnership*, the participants actually own a portion of a single aircraft. They pay a monthly charge based upon the actual cost of upkeep for the aircraft (hangar, insurance, reserve for overhaul, etc.). If an unexpected expense occurs, the members chip in equally to handle the bill. All decisions are made cooperatively by the owners, which generally number four or fewer. The only way out of the arrangement is to sell their portion of the plane, usually under some restrictions placed by the partnership agreement.

In a *flying club*, the organization owns the aircraft. Membership is achieved by paying an initiation fee, which might be refundable upon leaving the club. Decisions are made by an elected board of directors. Members pay a fixed monthly fee. The ratio of participants-to-aircraft is usually higher (one established club aims for 10 members per plane), and the club might own multiple aircraft.

Either way has advantages and disadvantages. With a partnership, you can point at the plane and say, "That's mine." But a major maintenance bill is also yours, at least part of it.

Individual expenses will be far less in a club, yet the pride of ownership isn't quite there. If you're leaving the FBO to escape scheduling problems, a club might not offer much relief.

Much of the following information applies to both arrangements. TAB's *Fly for Less: Flying Clubs and Aircraft Partnerships*, by Geza Szurovy, is an excellent in-depth guide for those contemplating joint ownership of either sort. The following sections apply to partnerships, although many of the details apply to flying clubs as well.

Member friction

Obviously, finding the right person(s) to collectively buy an airplane is paramount to the success of the endeavor. There are potentially a number of areas of friction.

Fiscal irresponsibility. Bills must be paid; paying them depends upon the participation of all parties concerned. If your partner gets behind on his portion of the expenses, you'll end up shouldering the burden. If the partner handles the financial side and neglects to make aircraft payments, that's your airplane that gets possessed.

It's more than just dealing with someone who won't pay a portion of the regular bills. You might feel it's time for the partnership to dig down and get the engine overhauled, but your partner might feel quite comfortable with the current state of the engine. Or your partner wants to get her instrument ticket and feels the panel should be upgraded. Yet all you want is a simple VFR fun machine.

Workload. Owning and maintaining airplanes is a lot of work. A partnership setup can reduce the impact on you, assuming the others provide their fair share. Like the monetary aspect, the load might end up in your court.

Pilot irresponsibility. You're driving along a country road, and you see the 172 that you own with three other partners. Suddenly, the Cessna's nose dips, then pulls up into a broad loop. What are you going to do about it?

Or perhaps you show up at the airport to go flying, just as your partner taxis in. You take off, but upon landing you find the FAA waiting. Seems a plane buzzed a local beach, and someone wrote down your N-number. You show the FAA the notebook where each member is supposed to record flight time, only to discover that the earlier pilot hadn't made any notations. You're the only one, officially, who had flown that day. What are you going to do next?

Incompatibility. I can tell when another member has flown our club home-built because the tiedowns are attached slightly differently. I think that my way is better, but I don't make an issue of it. What if we disagreed on something a little more vital, such as engine operation? How could we resolve it?

The partnership agreement

One way to avoid friction is to spell out the responsibilities and duties of each partner in a formal partnership agreement. Joint ownership of something as complex and valuable as an airplane should not be taken lightly. The partnership agreement should indicate the obligations and privileges of each member, and the actions to be taken during certain circumstances.

Some of the aspects that should be mentioned:

- How ownership expenses are assessed from the partners.
- Who handles the finances, and what is the spending limit before consulting the other partners?
- Restrictions on use (commercial, training, etc.).
- Responsibility for paying the insurance deductible in the event of an accident.
- Process by which one partner can buy out the other(s).
- How the airplane is scheduled.
- Arbitration procedures.
- Insurance to be carried.

Any number of other aspects should be addressed. I recommend that you don't try to assemble an agreement from scratch. Assemble several sample agreements and come up with one directly tailored to your situation. Szurovy's *Fly for Less* contains one such sample. In addition, AOPA's membership services department can provide a packet on co-ownership that includes another sample agreement.

LEASEBACKS

Ever wonder where those planes at the FBO come from? The FBO itself might own a number of them, but others are probably owned by private individuals. The individuals buy and maintain them, and the FBO rents them. In return, the owner gets a portion of the rental fee and preferential treatment for scheduling the aircraft for their own flying.

Leaseback arrangements are designed to benefit all parties. The FBO gets the use of an aircraft without the capital outlay of buying it. Pilots can fly an aircraft without the necessity of buying one. And the owner's expenses are alleviated by the rental fees paid by the pilots: *alleviated*, not necessarily *covered*.

Significant tax breaks used to be available for an owner who put an airplane on leaseback. These could convert a money-losing arrangement into one that at least broke even. Unfortunately, most of those benefits were lost during the tax reforms of the early 1980s. Some advantage, taxwise, is still retained. When carefully set up, a leaseback will allow you to deduct the expenses involved in owning your aircraft.

Leasebacks can work out for an aircraft owner. Unfortunately, there are tons of variables that you might not have control over. You can specify the minimum level of pilot qualification for renting your plane, but there's no way to control how gently (or otherwise) renters treat it. Does the leaseback agreement require you to have all inspections and maintenance done in the FBO's high-priced shop? How honest is the FBO? How stable is the FBO? What if you get committed to high monthly loan payments and the FBO goes bankrupt?

One nonvariable is significant: You'll still be responsible for all the expenses of the aircraft. Your plane will require 100-hour inspections, which are practically identical to the annual. Insurance premiums will be five times (or more) higher. You'll probably get preferential treatment when scheduling your own plane, but in many ways, it'll be just like renting.

Do your research. Talk to other owners who leaseback to the same FBO. Pick a popular aircraft, one that'll get a lot of rental time. Leasebacks aren't something to leap into. Much, but not all, of the feedback I got on leasebacks was negative. Yet some owners seem to make it work. Here's one who has.

CASE STUDY:
A SUCCESSFUL LEASEBACK

John Stephens wanted to fly since he was a teenager, but his problem was one many of us face: never having the required time and money. He finally said "the heck with it" and started lessons. Two-thirds of the way through his student training, he caught the buying bug. He knew owning was expensive and decided to try a leaseback. He asked his FBO for details and learned that the owner of one of their leased-back aircraft had put his plane up for sale. The airplane was an IFR-equipped 1981 Cessna 172P. Stephens felt a 172 (Fig. 2-10) was a good choice and was priced about right at $38,000.

He entered the leaseback arena in a businesslike manner. The money came from the sale of stocks he'd inherited. Instead of buying the plane outright, he formed a corporation. Stephens loaned the corporation $25,000 in return for stock. To finance the rest of the airplane, the corporation added a $10,000 long-term loan and a $5,000 short-term loan.

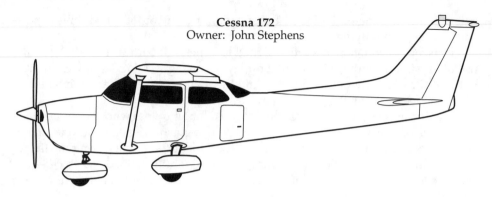

Cessna 172
Owner: John Stephens

Tiedown: $99/month
Insurance (full): $3,680/year*
Hours/year: 685*

Yearly inspections: $4,800*
Other yearly maintenance: $3,200*

*Aircraft on leaseback, which requires higher
insurance, plus 100-hour inspections in addition to
annuals. Owner receives portion of rental income from airplane.

Fig. 2-10. A summary of John Stephens' ownership expenses for his leased-back 172.

The long-term loan doesn't have a fixed payment schedule. Rather, the corporation pays interest until the loan is repaid. Over the four years that he's owned N51078, he's paid back about $8,000, with interest.

"Ownership is a personal decision," he says, "not just based on economic rationale, but also on convenience, pride of ownership, and affordability; however, just because something is affordable does not automatically make it a good investment."

Ownership experience

John's airplane is a good pick for a leaseback; just about everyone's willing to rent a nice 172. The plane has averaged about 685 hours per year. Almost 600 of those hours have been flown by renters.

The FBO manages the rentals and schedules the plane. They perform all the maintenance, but it isn't required by his leaseback agreement. All expenses are John's responsibility. The FBO collects the rental, keeping $10–$12 per hour as a management fee.

His agreement lets him "bump" anyone with 72 hours notice. He's exercised his right several times, but has never done so in a way that inconvenienced anyone. There has always been another plane for the renter to move to. He's used the airplane to get both his private license and instrument rating, and just recently received his commercial ticket. He's averaged one long trip a year, coupled with business trips in the 150–300 mile range.

Because the plane is in commercial service, Cessna Zero-Seven-Eight must receive 100-hour inspections. At the plane's high utilization rate, it undergoes an "annual" every two months or so; cost has ranged from $600–$1,500.

Other than "the usual avionics repairs," Stephens' 172 has required very little unexpected maintenance. Because it's a rental aircraft, he must replace a lot of tires and brake pads. "Also," he explains, "I take care of a lot of 'minor stuff' right away because the plane is rented out: things I normally could live with until some time later. This is partially pride, but also good business sense, as well as advisable from a legal liability aspect."

Total maintenance expenses run about $8,000 per year. Since he bought the aircraft, its Lycoming O-320-D2J has required one major overhaul; it cost about $9,000. Five dollars per flight hour are kept as a reserve for the next overhaul.

He has a prime tiedown location, a hard pad on the grass in close proximity to the FBO. Like all other expenses of the aircraft, the $99 per month tiedown fee is Stephens' responsibility. According to the tiedown agreement with the airport manager, he's also expected to supply his own ropes and mow the grass around his plane, but the FBO supplies these services at no cost.

He's insured via the FBO's fleet policy, at a savings over an individual policy. His portion of the premium is $3,680 per year. The FBO automatically deducts his share from the monthly rentals.

Advice

John Stephens is strongly in favor of aircraft ownership. "It makes flying easier, much more convenient, and cheaper if you fly a lot or enter into a leaseback and/or partnership. It's much easier to keep current, legally and practically, if you don't have to scrounge rental dollars at the last minute. Ownership at least lets you spread the costs out over time, without paying credit-card interest rates!"

As far as buying goes: "Proceed cautiously, and don't be in a rush to buy the first aircraft you like the looks of. Getting to know an A&P will help when it comes time to do a prepurchase inspection; don't neglect the inspection because it's the only thing that'll save you from a world of hurt if the plane isn't up to snuff."

Stephens' businesslike approach to his leaseback is probably one reason for his success. He analyzed the costs and set up a corporation to handle ownership, which makes it easier to deal with the IRS. But the business relationship and leaseback agreement with the FBO is the most important aspect.

"Not all leasebacks are the same, even at the same FBO. It pays to negotiate things that don't seem equitable. Talk to leaseback owners at several FBOs. Find out what they do and don't like about it. The key is having a good FBO, one that takes good care of the plane, is reputable, has good maintenance facilities, and that has enough business to give you the rental income that you need to be able to afford the plane," Stephens said.

"Also, ask to see the actual rental history for similar planes at the FBO and other FBOs at the same airport. Make sure you're not adding a plane to an already-saturated FBO. Try to take over an empty leaseback slot by buying a plane that has been on leaseback for some time and has a visible revenue and maintenance history.

"My experience has been very good," John concludes. "Much better than many that I hear about. For some like me, leaseback is the difference between having a plane or not. So far, I'm happier with leaseback than I think I would be in a partnership or flying club situation."

3

Picking the right plane

WHAT KIND OF PLANE are you going to buy? Of all the aspects of the aircraft-buying process, deciding *what* to buy is probably the most fun. Cessna? Piper? Mooney? Beech? New? Used? Taildragger? Trigear? Now we're talking about *airplanes*, instead of dry financial details.

Chapter 2 touched on parallels between airplane and car ownership. When it comes down to picking your aircraft, car selection experience doesn't count for much. After all, unless we're looking to buy a cheap second or third car, we're likely to just look at vehicles of relatively recent manufacture. Unless there are one or two accessories you must have, the cars' level of equipment won't make much difference to you. Unless you're a real hot-rodder or plan on heavy-duty towing, the performance differences won't be significant. At 55 mph, one car drives about like another. And no matter what kind of car you buy, it'll probably cost just about the same to own.

It's not the same in the flying world. To start with, the average general aviation airplane is about 22 years old. You probably wouldn't even consider a car that old. Yet half the aircraft in the United States are, by auto standards, jalopies.

Equipment? The age and level of an aircraft's avionics are crucial to the value of the plane. When AOPA had their "Like New 172" project, they bought the basic airplane for $22,000. They put a phenomenal $20,000 worth of avionics into the plane as part of the restoration process. Sure, they ended up with a well-equipped plane that was better than most of us probably need, but it indicates the significance of avionics to the value of a used plane.

Finally, there's performance. Most of us don't really care how fast a car can pull out of a parking space. Yet a plane's short-field performance can be an important part of our selection criteria.

So we've established that experience in the car world isn't going to help much in your search for the right kind of airplane. Let's take a look at determining what would be the right airplane for you.

WHAT'S YOUR MISSION?

It's easy enough to pick an airplane; there's the "eenie, meenie, miney, moe" method, for example. Or you can thumb through a magazine or *Janes' All the World's Aircraft* until you find a machine that catches your eye. Too often, though, an airplane is picked for the wrong reason. A buyer thinks that he needs a fast airplane and chooses just on the basis of speed, only to discover that the aeronautical hot rod can't handle the backcountry fields where he wants to pitch his tent.

The result is an unhappy camper. On occasion, the frustration can lead to a crumpled mass of aluminum when the owner tries to shoehorn the airplane into the wrong field. To avoid such problems, first determine your basic mission, then derive the performance and other requirements.

The mission statement

There are as many "missions" as there are pilots. Some pilots just like to fly for the pure joy of it, which is sometimes called "cutting holes in the sky." These folks could get by with a small, simple aircraft. The Aeronca Champ is a famous example; they don't call it the "Airknocker" for nothing. For those who prefer metal skins and tricycle gear, the Cessna 150 can't really be beaten for an inexpensive, simple aircraft.

Such basic planes aren't enough for some. They want room for radios and baggage, higher cruise speeds, and longer range. Some plan to operate from short grass fields, but others intend to use their planes for serious travel. "Serious" is not meant to imply that the travel is for business purposes; serious means that the airplane is used as a mode of transportation rather than pure entertainment. If it's more fun to take the Piper than the Pontiac, that's pure gravy.

Other needs exist. Perhaps you're interested in aerobatics, and a nice Decathlon (Fig. 3-1) seems like the way to go. Maybe you've got a yen for history that an antique or classic can scratch. Or maybe a floatplane or amphibian would be more your style.

But is the mission really right for you? Will you outgrow the simple little "puddle jumper"? Are you really that interested in aerobatics? Do some serious thinking about how your needs might change.

Sometimes it's tough to determine a single mission. For example, you'd like a fun little knockaround aircraft, but still require fast cross-country travel. Keep in mind that good cross-country aircraft can be rented at practically any FBO. Buy the little J-3 for fun, and rent a Piper Arrow or equivalent for heavy-duty traveling.

To start, come up with a simple sentence or phrase expressing your basic mission: "Carry one passenger on long cross-countries between major airports." "Amateur competition aerobatics." "Operate from short private airstrip, with high-speed travel to nearby destinations." "Good trainer for kid's lessons." Determining your mission might seem trivial, yet it is an important step in figuring out which plane fits your requirements for the least money.

Fig. 3-1. The American Champion Decathlon is one of the few aerobatic airplanes on the market.

The mission pitfall

You've determined your mission and are ready to start scanning the performance figures. Before you do, take another look: Is your stated mission *really* what you need? Because if you buy *more airplane than necessary*, not only are you wasting money, but you risk reducing your flying enjoyment.

Say, for instance, that your first cut at a mission statement was "fly two adults and two kids on long cross-countries." That's pretty reasonable. You'll obviously need a four-seat airplane with a moderate amount of fuel capacity. Any number of planes would do; the Cessna 172 and Piper Warrior come to mind.

However, you know that a 172 with four people on board (even if two of them are kids) can get somewhat marginal on useful load. It's tempting to modify the mission statement: "To fly two adults, two kids, and 200 hundred pounds of baggage on long cross-countries." Oh-oh, that means you'll need a higher useful load. The plane has just moved into the Cessna 182 and Piper Archer category (Fig. 3-2).

Then you think, "Hmm, I might want to take the folks along." Up goes the mission statement to "Fly four adults and two kids . . .." And you've just kicked the plane into the Cessna 207 and Cherokee Six category.

What's wrong with that? Nothing, if you really need the capacity. What is it costing? Money, of course. That Cherokee Six will probably cost $15,000 more than the Warrior you started with. It'll burn nearly twice the fuel and have a proportionately higher maintenance cost.

Piper Aircraft Corporation

Fig. 3-2. The Piper Archer can actually haul four adults and baggage in its four seats, unlike many smaller "four-seat" aircraft.

Plus, if you think you'll be doing most of your flying solo, the plane will be less pleasant to fly. Those load-haulers have a pretty broad CG range. When flown solo, they'll be at the forward limits of their CGs. You'll have to horse back on the yoke to rotate, and you might not have enough elevator authority to keep from landing nosewheel-first. Handling a large airplane might rob you of some of the fun of ownership.

You have to draw the line somewhere; otherwise you could end up with a DC-3 instead of a Cessna 180. Look at your mission realistically, and try to look at those airplanes that give just the right performance.

Ownership without guilt

"Mission?" some are wondering. "I just want to own a Hotrock 200!" In other words, you're not making a logical, rational decision based upon careful analysis of needs and features. There's nothing wrong with that!

Few of us *have* to own an airplane. We've already established that ownership is a hobby; the airplane is not expected to earn its own way; however, there are things you expect to do with your Hotrock 200. Maybe your fantasy is flitting to that little mountain strip near your favorite fishing hole, or packing your family of four aboard and winging your way to Grandma's at 200 knots.

Before you spring the bucks, shouldn't you at least make sure the plane matches your needs? One rite of passage to adulthood is buying your first car.

44   *Picking the right plane*

Considerable effort goes toward finding exactly the right set of wheels to fit your self-image. When found, it is bought, whether Mom and Dad approve or not.

Reality soon intrudes. The engine leaks oil like a grounded tanker. The chassis is sprung, the shocks are shot, and the electrical system shorts out. Have you ever seen a shiny older car at the side of the road with a stricken teenager tenderly stroking the steering wheel while waiting for a parent and a tow rope? Too often, that first car is one's introduction to the need for rational decisions, instead of emotional ones.

It's bad enough with a $500 car. Do you want to go through the same process with a $25,000 airplane? Shouldn't you make sure the plane is what you need and not just what you want? But hey, it's your money and your decision. If your analysis finds that Brand X better meets your needs than the Hotrock 2000, but you still prefer the Hotrock, go for it!

PERFORMANCE AND FEATURES

It's all well and good to make a highfalutin statement regarding what your airplane's going to be used for. Let's break the simple mission statement into hard requirements.

Performance requirements

First, let's look at how much runway you have. If your mission calls for operation from "standard" airports only, there's no real takeoff-and-landing requirement; a standard airport has 3,000–4,000-foot runways, fairly clear approaches, and the like.

When you're off the ground, four interrelated parameters are cruise speed, range, fuel capacity, and fuel consumption. Because refueling operations severely impact total time en route, set the endurance requirement first (usable fuel divided by fuel consumption at cruise). Some planes have only two or three hours of onboard fuel, which is fine for local pleasure flights, but serious cross-country travel requires more. Again, set the minimum value based on your mission. Be realistic; how long do you honestly want to fly without a restroom pit stop? A typical value might be 3 hours, add another 30–45 minutes for reserves, and round it up to 4 hours.

Now state your desired cruise speed. Local sport flying, of course, doesn't need a cruise speed requirement. In any case, take your endurance requirement from the last paragraph and divide by cruise speed to determine your aircraft's required range.

If your requirements include operation from smaller strips, your selection process just became more difficult. Published performance figures don't really indicate which airplanes are truly better at short-field work. For instance, the American Yankee and the Beech Bonanza list takeoff distances of about 900 feet, yet the Yankee has proven to be a poor short-field performer in comparison.

You'll have to dig a bit deeper for information if you'll be flying from shorter runways. Talk to owners, and read some of the magazines and books that review various aircraft.

Features

Performance isn't the only issue. Your mission might require that the aircraft have certain features. Not to mention catering to your own biases. What's your preferred configuration? High wing? Low wing? Side-by-side seating? Tandem? There are advantages and disadvantages to each.

Generally, a low-winger gives better flight visibility for maneuvering in crowded airspace. A high-wing aircraft is better for sightseeing.

Need IFR capability? You can turn just about any airplane into an IFR platform with enough money and a willingness to forego common sense. I've seen pictures of full-IFR Super Cubs and Waco biplanes. Yet there's more to it than mere panel space. Everyone agrees that the superior stability of a Piper Warrior makes it a better platform for instrument work than a Luscombe, for example.

Other panel questions: Flying in and around congested airspace? Mission call for long cross-countries? Multiple communication radios will be handy; put GPS on the wish list.

Where will you keep the aircraft? An enclosed hangar widens the options. You could opt for a fabric-covered bird, or a wood-winged plane like a Bellanca or early Mooney. Otherwise, you should shoot for all-metal construction.

How many seats? It's an unfortunate aviation fact of life that most planes can't legally, safely, and/or comfortably haul full fuel plus as many people as they have seats for. A Cessna 120 is not the platform for two large people, especially if they plan to take luggage. In many circles, a 172 is considered a three-seat airplane, rather than a four-seat airplane.

In any case, it's nice to have some margin. If you're at a toss-up between a two- or four-seater, the larger aircraft is usually the better way to go. It'll give you the most flexibility, and it is quicker to sell when you put it on the market. A four-seater is not much more expensive to operate, either.

However, two-seaters are invariably cheaper to buy. If you do decide that a two-seater will fit your mission, be aware that the seat configuration plays a role in performance and pilot/passenger satisfaction. A tandem-seater aircraft, which has one seat in front and the second in back, like a Cub or Champ, gives the pilot good visibility to either side and places him on the centerline. Because the aircraft can be narrower (only one person wide), tandem-seaters are theoretically faster. Tandem-seaters are usually more comfortable to ride in because each occupant gets at least 24 inches of shoulder room rather than sharing 40 inches with the passenger; however, communication between occupants requires strong lungs or an intercom.

Side-by-side aircraft are more civilized. Both occupants can see the instruments and easily bellow into their corider's ear. Because aircraft engines are as wide as two people anyway, the performance cost is often negligible. If you've got a husband who's not especially thrilled with flying, he might be more comfortable sitting alongside you rather than stuffed in the back of the plane. (Twenty years after I gave my mother her first and only lightplane ride, she told me that she had been claustrophobic in the back of the Citabria because she couldn't see forward.)

One last configuration consideration is the landing gear: conventional (tail-dragger) or tricycle gear. Conventional gear is more rugged and better suited to rough and short fields. A tailwheel is usually less complex and is subjected to less strain than a nosewheel, which results in lower maintenance costs.

Again, conventional-geared aircraft are theoretically faster than trigeared types because tailwheels create less drag than nosewheels. Most pilots are more comfortable with tricycle gear. There's no question that trigears are easier to operate and are safer. A taildragger checkout is sometimes difficult to arrange, too. Don't let taildraggers scare you off. You don't have to be Chuck Yeager to taxi and fly a Champ.

Performance and features summary

As of this point, you should know the approximate performance figures for your dream airplane, as well as the preferred configuration and features. Use these parameters to narrow down the list of potential candidates.

Picking the right plane

You have decided what the airplane will do. How do you pick the right plane to do it? Based upon previous decisions, you should have a prepared list with the following requirements:

- Endurance.
- Cruise speed.
- Number of seats and seating configuration (if two-seater).
- Avionics level (basic VFR, IFR, GPS, etc.).
- Construction (all-metal to allow outside storage, etc.).
- Landing-gear configuration.
- Any special performance requirements (short field, etc.).

The real world

The real world is sometimes hard to pin down, and general aviation aircraft are no exception. Take the lowly Cessna 172 (Fig. 3-3), for instance. It was in production from 1957 until 1984. Easy enough, you might think: "All I have to do is pick the appropriate year where the average prices correspond with the money that I have available."

However, Cessna didn't stand still during 27 years of production. The 172 switched from a Continental to a Lycoming for 1968 and to a slightly bigger Lycoming 10 years later. The gross weight was gradually increased from 2,200 to 2,400 pounds. The rate of climb started at 660 fpm, rose to 770 fpm by the late 1970s, and dropped to 680 fpm by the time production ceased.

But at least it was built by a single company, under a single name. You're probably familiar with the lowly Ercoupe, a funky-looking twin-tailed cutie from the 1940s that offers surprising performance on a 90-horse engine. If you were in the

Fig. 3-3. The Cessna 172 is probably the most popular light aircraft of all time. Yet not all models of the 172 are alike.

market for one of these, you'd be looking for a plane manufactured by companies like Ercoupe, Forney, Alon, and Mooney. By the time Mooney got hold of it, the twin tail was gone and the nameplate said Mooney Cadet.

You have your job cut out for you. But hey, picking out what kind of plane to buy is the fun part, remember? The remainder of this chapter examines some of the issues involved in selection. To get you started, Table 3-1 contains specifications and information for a number of single-engine aircraft.

Now, this is by no means a complete list. It's just a sample of some of the airplanes available. I've tried to pick out planes that had longer production runs or evolved into other airplanes. Undoubtedly, I've left off someone's favorite—sorry about that.

The performance figures included on the table were those published by the manufacturer at the time the plane was built. There is a considerable amount of optimism evident; I'm afraid the marketing departments were just as responsible for these numbers as the flight test section. I really chuckled when I learned that Cessna claimed that my old 150 cruised at almost 120 mph. Then again, 30-year-old airplanes are not going to perform at brand-new specifications. In other words, take the table as a guideline, not gospel.

Here's an explanation for the columns on the table:

Manufacturer. The name of the company that first built the model of aircraft; other companies that subsequently made the same plane are listed in parentheses underneath the original manufacturer. For instance, look under Ercoupe. You'll see the specifications for the original Ercoupe Model E, Ercoupes built by Forney, and the modified version sold by Mooney.

Model. The model name/number that the aircraft was sold under, as well as its "street" name. For instance, the 172 and Skyhawk are two different models to a Cessna purist; the Skyhawk featured upgraded avionics and more standard equipment as compared to the 172. The table usually gives the performance values for the fancy version.

Table 3-1. Used aircraft specifications.

Manufacturer	Model	Year	Engine Make	HP	Gross	Useful Load	Cruise	Stall	Range	Seats/Config.	Gear
			All Figures Miles per Hour and Statute Miles								
Aeronca	11CC Super Chief	49	Continental	65	1350	530	95	45	420	2 SBS	Conv.
	7AC Champ	50	Continental	90	1450	560	100	44	350	2 Tan	Conv.
(Bellanca)	7ACA Champ	72	Franklin	65	1220	470	86	28	310	2 Tan	Conv.
	15AC Sedan	53	Continental	145	2050	900	114	53	450	4	Conv.
(Champion)	7EC Champ	57	Continental	90	1450	630	100	40	410	2 Tan	Conv.
(Champion)	7ECA Citabria	70	Lycoming	108	1650	670	112	50	450	2 Tan	Conv.
(Champion)	8KCAB Decathlon	70	Lycoming	150	1800	570	125	53	565	2 Tan	Conv.
(Bellanca)	8GCBC Scout	75	Lycoming	180	2150	835	125	52	575	2 Tan	Conv.
American	AA-1 Yankee	70	Continental	108	1500	550	134	58	520	2 SBS	Tri
(Grumman)	AA-1B	75	Lycoming	108	1560	580	124	60	490	2 SBS	Tri
(Grumman)	TR-2	75	Lycoming	108	1560	525	127	60	510	2 SBS	Tri
(Grumman)	AA-5A Traveler	75	Lycoming	150	2200	950	147	58	680	4	Tri
(Grumman)	AA-5B Tiger	75	Lycoming	180	2400	1115	160	61	710	4	Tri
(American General)	AG-5B Tiger	93	Lycoming	180	2400	990	165	61	630	4	Tri
Aviat	Husky	93	Lycoming	180	1800	600	140	42	633	2 Tan	Conv.
Beech	Bonanza B35	50	Continental	165	2650	1060	170	56	750	4	Tri - Retract
	Bonanza K35	60	Continental	250	2950	1120	200	60	910	4	Tri - Retract
	Bonanza V35B	70	Continental	285	3400	1430	203	63	600/1110	4 - 6	Tri - Retract
	Bonanza V35B	80	Continental	285	3400	1280	198	59	825	4 - 6	Tri - Retract
	Bonanza F33A	90	Continental	285	3400	1170	198	59	825	4 - 5	Tri - Retract
	A36	80	Continental	285	3600	1410	193	60	800	4 - 6	Tri - Retract
	A36	90	Continental	300	3650	1400	202	68	1000	4 - 6	Tri - Retract
	Musketeer	65	Continental	165	2350	925	137	59	780	4	Tri
	Musketeer Sport	70	Lycoming	150	2250	900	131	56	760	4	Tri
	Sundowner 180	80	Lycoming	180	2450	950	120	59	690	4	Tri
	Sierra	80	Lycoming	200	2750	1040	158	69	750	4	Tri - Retract
	Skipper	80	Lycoming	115	1680	580	110	54	380	2 SBS	Tri
Bellanca	Cruiseair	50	Franklin	150	2150	900	165	44	680	4	Conv. - Retract
	Cruisemaster	50	Lycoming	190	2600	1070	180	41	680/1000	4	Conv. - Retract
	260C	70	Lycoming	260	3000	1150	186	62	1000	4	Tri
	Viking 300	70	Lycoming	300	3200	1250	194	70	825	4	Tri - Retract
Cessna	120	49	Continental	85	1450	660	100	41	450	2 SBS	Conv.
	140	50	Continental	85	1500	600	105	44	450	2 SBS	Conv.
	150	59	Continental	100	1500	540	121	48	520	2 SBS	Tri

Continued on page 50

Table 3-1. Continued.

Manufacturer	Model	Year	Engine Make	HP	Gross	Useful Load	Cruise	Stall	Range	Seats/Config.	Gear
	150	65	Continental	100	1600	630	122	50	490	2 SBS	Tri
	150	70	Continental	100	1600	625	117	48	450/725	2 SBS	Tri
	152	80	Lycoming	110	1670	530	123	50	370/630	2 SBS	Tri
	170	50	Continental	145	2200	1020	120	52	480	4	Conv.
	172 Skyhawk	57	Continental	145	2200	940	124	52	550	4	Tri
	172 Skyhawk	60	Continental	145	2200	940	124	52	620	4	Tri
	172 Skyhawk	65	Continental	145	2200	940	120	52	620	4	Tri
	172 Skyhawk	70	Lycoming	150	2300	1050	131	49	620/640	4	Tri
	172 Skyhawk	80	Lycoming	160	2400	980	138	53	500	4	Tri
	175 Skylark	60	Continental	175	2350	1040	139	52	595	4	Tri
	177 Cardinal	70	Lycoming	180	2500	1140	138	57	650	4	Tri
	Cutlass RG	80	Lycoming	180	2650	1100	161	58	830	4	Tri - Retract
	182 Skylane	60	Continental	230	2650	1120	157	57	660	4	Tri
	182 Skylane	70	Continental	230	2800	1160	160	57	690/910	4 - 6	Tri
	182 Skylane	80	Continental	230	2950	1310	164	55	900	4 - 6	Tri
	182 Skylane RG	80	Lycoming	235	3110	1300	180	57	970	4 - 6	Tri - Retract
	190	50	Continental	240	3350	1330	160	60	750	4	Conv.
	195	50	Jacobs	300	3350	1320	165	60	750	4	Conv.
	207 Skywagon	70	Continental	300	3800	1920	158	67	585/775	7	Tri
	207 Stationair	80	Continental	300	3800	1700	165	67	400/600	8	Tri
	210 Centurion	60	Continental	260	2900	1160	190	59	755	4	Tri - Retract
	210 Centurion	70	Continental	285	3800	1720	188	65	765/1065	6	Tri - Retract
	210 Centurion	80	Continental	300	3800	1600	197	65	835	6	Tri - Retract
Ercoupe	Ercoupe Model E	50	Continental	85	1400	560	110	50	430	2 SBS	Tri
(Forney)	Ercoupe	60	Continental	90	1400	510	121	50	500	2 SBS	Tri
(Alon)	Ercoupe	66	Continental	90	1450	520	124	50	450	2 SBS	Tri
(Mooney)	Cadet	70	Continental	90	1450	500	118	46	520	2 SBS	Tri
Interstate	Arctic Tern	80	Lycoming	150	1900	830	117	34	550	2 SBS	Conv.
(Arctic)	Arctic Tern	90	Lycoming	150	1900	830	117	34	550	2 SBS	Conv.
Lake	LA4	70	Lycoming	180	2400	800	131	51	625	4	Tri - Retract
	LA4-200	90	Lycoming	200	2690	1030	150	45	650	4	Tri - Retract
	Renegade	90	Lycoming	250	3050	1200	152	56	1030	6	Tri - Retract
Luscombe	8F Silvaire	49	Continental	90	1400	590	115	48	450	2 SBS	Conv.
Maule	M4 Jetasen	70	Continental	145	2100	1000	150	40	700	4	Conv.
	M4 Rocket	70	Continental	210	2300	1080	165	40	680	4	Conv.
	M5 180C	80	Lycoming	180	2300	975	127	56	490/750	4	Conv.
	M-6 Super Rocket	90	Lycoming	235	2500	1000	160	35	490/860	4	Conv.
Meyers	125C	50	Continental	125	1675	580	142		500	2 SBS	Conv.
(Aero Commander)	200	60	Continental	240	3000	1130	195	55	700/1300	4	Tri - Retract
Mooney	Mark 20	57	Continental	285	3000	1060	218	54	1230	4	Tri - Retract
	Mark 20A	60	Lycoming	150	2450	1035	165	57	750	4	Tri - Retract
	M20C Ranger	70	Lycoming	180	2575	970	180	57	630/900	4	Tri - Retract
	M20E Chaparral	70	Lycoming	200	2525	1050	176	57	1000	4	Tri - Retract
	M20F Executive	70	Lycoming	200	2750	940	174	61	980	4	Tri - Retract
	201	80	Lycoming	200	2740	1110	185	62	1150	4	Tri - Retract
	201SE	90	Lycoming	200	2750	1100	195	64	1120	4	Tri - Retract
						970	193	61	1095	4	Tri - Retract

Table 3-1. Continued.

Manufacturer	Model	Year	Engine Make	HP	Gross	Useful Load	Cruise	Stall	Range	Seats/Config.	Gear
Morrisey	2150	60	Lycoming	150	1817	690	135	52	525	2 Tan	Conv.
Piper	PA-18 Super Cub 95	50	Continental	95	1500	710	100	38	350	2 Tan	Conv.
	PA-18 Super Cub 105	50	Lycoming	108	1500	675	105	38	350	2 Tan	Conv.
	PA-18 Super Cub 150	60	Lycoming	150	1750	820	115	38	460	2 Tan	Conv.
	PA-18 Super Cub 150	70	Lycoming	150	1750	820	115	43	460	2 Tan	Conv.
	PA-20 Pacer	50	Lycoming	125	1800	820	125	48	535/655	4	Conv.
	PA-22 Tri-Pacer	60	Lycoming	160	2000	890	134	49	535/655	4	Tri
	PA-24 Comanche	60	Lycoming	180	2550	1075	160	58	800	4	Tri - Retract
	PA-24 Comanche	60	Lycoming	250	2800	1200	181	64	780	4	Tri - Retract
	PA-24 Commanche C	70	Lycoming	260	3200	1430	185	65	735/1130	4	Tri - Retract
	PA-28-140C Cherokee 140	70	Lycoming	150	2150	940	133	55	525/725	2 - 4	Tri
	PA-28-180 Cherokee 180E	70	Lycoming	180	2400	1100	143	57	725	4	Tri
	PA-28-161 Warrior	80	Lycoming	160	2325	985	146	57	600	4	Tri
	PA-28-161 Warrior II	90	Lycoming	160	2440	1000	145	58	600	4	Tri
	PA-28-181 Archer	80	Lycoming	180	2550	1130	148	61	560	4	Tri
	PA-28-181 Archer II	90	Lycoming	180	2550	1040	148	61	600	4	Tri
	PA-28-200 Cherokee Arrow	70	Lycoming	180	2500	1080	162	64	860	4	Tri - Retract
	PA-28RT-201 Arrow IV	80	Lycoming	200	2750	1110	165	63	830	4	Tri - Retract
	PA-28-236 Dakota	80	Lycoming	235	3000	1400	166	64	750	4	Tri
	PA-32 Cherokee Six	70	Lycoming	300	3400	1685	168	63	525/880	6	Tri
	PA-32-301 Saratoga	80	Lycoming	300	3615	1695	173	69	860	6	Tri
	PA-38-112 Tomahawk	80	Lycoming	112	1670	570	124	56	520	2 SBS	Tri
Rockwell	Lark Commander	70	Lycoming	180	2475	940	132	59	560	4	Tri
(Commander)	Commander 112A	75	Lycoming	200	2650	960	161	62	975	4	Tri - Retract
	112B	93	Lycoming	260	3250	1200	189	65	775	4	Tri - Retract
Ryan	Navion	50	Continental	185	2750	970	155	54	500/750	4	Tri - Retract
Stinson	180 Voyager	49	Franklin	165	2400	1110	130	51	550	4	Tri
Taylorcraft	F-19	50	Continental	65	1200	450	95	38	350	2 SBS	Conv.
	F-22	90	Lycoming	118	1750	660	117	41	640	2 SBS	Conv.

Year. The year of aircraft for which the performance figures are given.

Engine make. The company that manufactured the engine.

HP. The engine's rated horsepower.

Gross. The allowable gross weight.

Useful load. The difference between the gross weight and the manufacturer's listed empty weight. It is a certainty that any aircraft you look at will have a lower useful load than that listed here.

Cruise. The claimed cruise speed (at 75 percent power) in statute miles per hour.

Stall. The stall speed in statute miles per hour.

Range. The published range of the aircraft when flown at 75 percent power. Where two values are given, the second one indicates the range when optional larger fuel tanks are installed. **Note:** The range figures are as released by the indicated manufacturers at the time the aircraft were built. The different companies often published different sorts of figures. Some published a range to dry tanks; others included a 30-minute VFR or 45-minute IFR reserve.

Seats/configuration. The number of passengers the plane can legally carry. If the aircraft is a two-seater, SBS stands for a side-by-side seating configuration, and "Tan" means tandem.

Gear. The type of landing gear. "Conv." means conventional (taildragger), and "Tri" means tricycle. "Retract" means the aircraft has retractable gear.

Now that we have some background on aircraft data, let's try a sample of the aircraft selection process.

A SAMPLE SELECTION

Let's say that your mission is something like: "Two-seat side-by-side airplane for fun flying and sightseeing. Must be inexpensive to own and operate." That is a pretty typical first-time-buyer's mission.

Specifying a two-seater narrows down the field. The "sightseeing" aspect kind of aims us toward a high-wing airplane, eliminating the Piper Tomahawk and the Ercoupe. Side-by-side seating cuts out planes such as Cubs and Champs. Looking at Table 3-1, you're left with planes such as the Aeronca Chief, Luscombe, Cessna 120/140/150/152 series, Arctic Tern, and the Taylorcraft.

You've also specified the need for inexpensive ownership. The cheapest storage option is an outside tiedown; that rules out most of the fabric-covered planes. What are we left with? The two-seat Cessnas (except the 120 and early 140s, which had a fabric covering on the wings) and the Luscombe. Luscombes are fine airplanes, but they are getting a bit rare, which has driven their prices up.

So now we've established that we're in the market for a nice Cessna 140, 150, or 152. They're really a good pick for a first-time aircraft. The line has good parts availability, and any A&P will be well familiar with their problems and foibles.

So should you just find the best one in your price range and buy it? You could work that way. The physical condition and reliability of the aircraft will have the most contribution to your satisfaction with ownership. Pick one that's mechanically sound and you'll probably be happy with it.

Still, wouldn't you like the most for your money? The 140/150/152 line was built for more than 30 years; you can get the features that fit you the best by picking the right year.

And that's true for any airplane, not just the little Cessnas. Look at the useful load column on Table 3-1. Older versions of several airplanes, not just the Cessna 150, have a higher useful load than newer models.

Let's see how some of the changes might affect your aircraft selection. We'll keep the two-seat Cessnas for an example. Even if you're interested in a larger, more complex aircraft, please follow along. Most aircraft show similar evolution over their design life.

External differences

Figure 3-4 shows the external changes in the 140/150/152 series over the years. The line started out as a rounded-tail conventional-gear aircraft. The 150 series started in 1959, with the major apparent change being the squared-off tail (to match the 172) and tricycle gear. The tunnel effect of the fuselage was relieved in 1964 with the addition of the cockpit's rear window; Cessna called it "Omni-vision."

With the addition of a swept tail, the '66 model was the first of the "modern" 150s. The '71 lengthened the dorsal fin; the '75 added a slightly taller tail. Finally, the tried-and-true Continental O-200 was replaced by the Lycoming O-235, resulting in a different-shaped cowl and a designation change to the Model 152.

Which looks better? You probably said the '66-models on because the swept-tail birds are more common. Still, did you ever wonder why the tail was swept? A 100-knot airplane didn't need it for speed. It was instigated by the Cessna marketing department purely for sleeker aesthetics. Did it produce an aerodynamic benefit? To the contrary, the rudder is actually *less effective* than that of the earlier, straight-tail models; the hinge line isn't perpendicular to the relative wind. The tail was raised 6 inches in 1975, gaining back some of the lost efficiency.

Don't be too hard on the Cessna marketeers. Sales of the 150 simply exploded with the '66 model. Aerodynamically wasted or not, the swept-tail 150s sold like gold-plated hotcakes.

Let's look back even further, before the rear window. It was also a marketing response, but one most pilots agree with. That extra Plexiglas area makes the cabin far airier and more pleasant. The straight-back earlier models look almost old-fashioned.

Then again, perhaps that's an advantage. One problem with owning a 150: People assume you're a student pilot. When you walk into an FBO, people ask, "Do you need someone to sign your logbook?" You won't have that problem with a straight-backed 150. It's a poor-person's classic airplane that is rare enough to be noticeable, yet as cheap to run as any other 150. The 140 is even better.

If you don't want to look like the crowd, you might set your sights on an earlier model, which is applicable to more aircraft than just the Cessna 150 series.

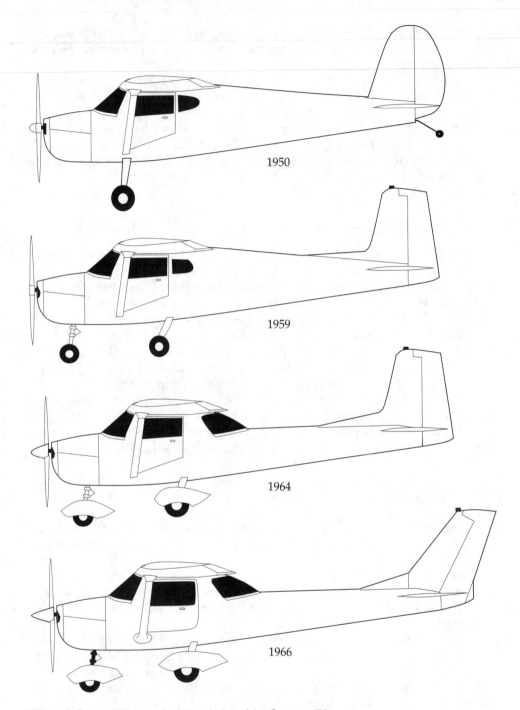

1950

1959

1964

1966

Fig. 3-4. External changes in the evolution of the Cessna 150.

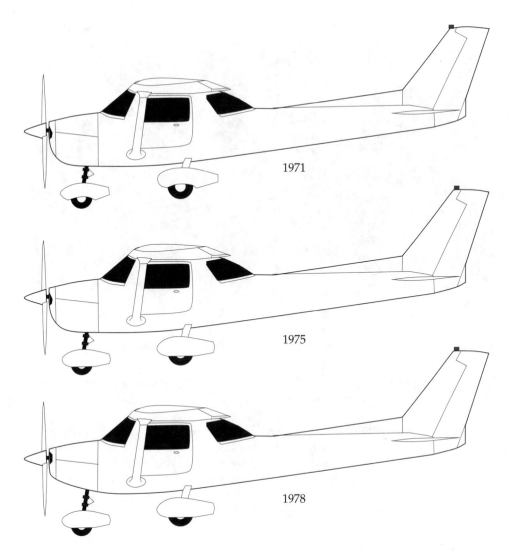

Fig. 3-4. Continued.

Landing gear

Easy enough to see the difference here, right? The 140 is a taildragger, and all the 150s and 152s are tricycle. Still, you might not want to lump all the 150s/152s together. There are subtle differences through the line, with several changes.

The first came in 1961, when the main gear was moved back 2 inches. The plane had a tendency to fall back on its tail if you fully loaded the baggage compartment before anyone got in. The gear was changed again 10 years later. The track of the main gear was increased to provide more stability.

Lack of stability, of course, is the problem with the Cessna 140's conventional gear. In addition to groundloops, some 140s had a tendency to nose-over as well.

Fig. 3-5. The noseover tendency of the Cessna 120/140 series brought about these aftermarket extenders. Note how they move the wheel a few inches forward of the gear leg.

Look at the gear legs of a typical 120 or 140; often you'll see a set of aftermarket extenders (Fig. 3-5) that move the main wheels forward a few inches to reduce this tendency.

Should you reject the 140 just because of the conventional gear? Of course not. All it takes is training, and a little more attention. During the 1950s and 1960s, a number of airplanes transitioned from taildragger to tricycle gear. The Cessna 170 became the immortal 172. The Cessna 180 became the 182. The Piper Pacer begat the Tri-Pacer. Because fewer buyers are interested in taildraggers, their prices sometimes are lower than their trigear equivalent.

Engine choices

Unless you've had a bad experience or two, you probably don't have any biases about engines. One thing to keep in mind: When you go outside the normal Lycoming and Continental arena, you're likely to encounter rare and/or high engine part prices. That should not be much of a problem with our target aircraft line because the 140 had an 85- or 90-horse Continental, and the 150 had the tried-and-true O-200 (100 horsepower).

The 152 is essentially a 150 with a Lycoming O-235 substituted for the Continental. Therein lies one problem. Cessna made the swap so that the 152 could run on 100-octane low-lead aviation fuel. The Continental was designed for 80-octane; even the lower lead content of 100LL fouled spark plugs or otherwise caused problems. Most of the major fuel refiners were eliminating 80-octane fuel, so Cessna felt bound to change.

A few years later the EAA ran an extensive set of flight tests and gained approval to use ordinary unleaded auto gas in the Cessna 150. At one swoop, the EAA reduced the direct cost of operating by almost one-half.

So, if you buy a Cessna 152, you'll be stuck with using high-priced aviation fuel. It is actually possible to modify the engine to gain autofuel approval, but high-priced engine parts must be replaced.

Using autofuel in your airplane is not a free ride. If you plan to fly a lot of cross-countries, you'll find 100LL a lot more available on most airports than 80-octane or autofuel. In this case, it would be better to own an airplane designed to use 100LL. But for the lowest cost of ownership, pick a plane for which STC approval is available.

Equipment and features

Just because a plane is older doesn't mean it contains out-of-date, obsolete avionics. A 1964 Cherokee probably has had its avionics upgraded since it was new; a 1976 model might still have its original suite. Then again, maybe that '64 model was upgraded in 1972.

There are other features to consider. Let's look back at the 150 series again. The original models had manually-activated flaps. A large lever (like a small car's emergency brake) was mounted between the seats. Pull the bar upward, and the flaps come down. The mechanism had notches at 10°, 20°, 30°, and 40°.

In 1966, Cessna added an electric motor to the flap system and eliminated the lever. That was an unfortunate decision, in my opinion. The mechanical system was a lot more reliable; the electric flaps were soon the subject of a recurring airworthiness directive. The mechanical flaps are a lot faster to operate, and you can tell the flap position by feel rather than having to glance at a panel indicator.

When Cessna transitioned to the 152, they made one more change, eliminating the 40° flap setting. Too many pilots were crashing during go-arounds by forgetting to raise the flaps. This was a pretty common change to the Cessna line about this time, so if you need a short-field bird, you might want to concentrate on earlier models.

Cessna made one other change to its lines starting around 1978: conversion to 24-volt electrical systems. I know the engineering advantages of the 24-volt bus; wire diameters can be reduced, and starters have a bit more ooomph.

Yet a 24-volt system is a 24-carat hassle as far as the private aircraft owner is concerned. Twenty-four volt batteries cost almost double. Ordinary parts such as bulbs and solenoids cost more and are far less available. You can buy a cheap 12-volt battery charger at any auto parts store; you'll have to get a 24-volt version from an aviation supplier. And if you're ready to fly and the aircraft's battery is dead, you can't jump-start it from your car. For ease of ownership, stick with planes with 12-volt systems.

What about age?

We've been talking a lot about buying older airplanes from the 1960s and even the 1940s and 1950s. Are there dangers involved with planes that old? What about corrosion? True, you'll have to look out, but corrosion isn't just an old-airplane phenomena. It can happen with relatively recent models, too, depending on where they're based and how they're kept.

Parts can be a problem with older planes. The Continental A-65 engine used in a lot of the small planes of the 1950s and 1960s has been out of production for more than 30 years. Several aftermarket companies have ensured that parts stay available for this and other out-of-production engines, although the prices continue to rise.

Similarly, companies have sprung up to supply replacement parts for the airframes as well. Some companies manufacture enough parts to practically build a compete plane, although the price would be a tad high. The additional maintenance involved with older airplanes is often balanced by the lower purchase price.

Where do you find out this stuff?

There's been a lot of information about the Cessna 150/152 series packed into the last couple of pages. Obviously, you'd like a similar level of detail for the type of plane you're interested in.

One of the best sources of information is *Aviation Consumer* magazine. They publish an in-depth analysis of a specific aircraft type in almost every issue. These articles include the differences between models, maintenance problems, safety record, and price history. The magazine is published bimonthly; subscriptions cost $84 annually. Write:

The Aviation Consumer
Belvoir Publications, Inc.
75 Holly Hill Lane
P.O. Box 2626
Greenwich, CT 06836-2626

Many of the magazine's aircraft analyses have been reprinted in *The Aviation Consumer Guide to Used Aircraft*, available through TAB Books. Another good source of information is Bill Clarke's *The Illustrated Buyer's Guide to Used Airplanes*, also available from TAB. Clarke discusses a number of aircraft from the owner's point of view and provides a summary of AD notes on each model. Books on specific aircraft lines are widely available (Fig. 3-6). There are several on the Cessna and Piper singles, and other volumes on specific models.

In addition, type clubs have sprung up around many aircraft models. These clubs can provide a flood of information; many offer hints for prospective buyers. A list of these clubs and associations can be found in appendix B.

CASE STUDY:
TRADING UP

What's the Cadillac of general aviation aircraft? Most pilots give one answer: The Beechcraft Bonanza. From its premiere in 1947, the Bonanza has set the standard for light aircraft. Fast, comfortable, and well built, Beechcraft's single-engine flag-

Fig. 3-6. Your local aviation bookstore will have dozens of books on various popular general aviation aircraft.

ship has been in continuous production for more than 45 years. What are they like to own? What do they cost to own?

Let's meet Claude Abbott (Fig. 3-7), an engineer with a major aerospace firm. Bonanza N11JQ (Fig. 3-8) wasn't Abbott's first airplane. He'd owned a 1947 Luscombe in the early 1970s. His second airplane, a 1977 Piper Warrior, became the key to his eventual step up to a Bonanza. He'd found the Piper in 1984 for $35,000. He sold his boat to get together a $12,000 down payment and applied to a local bank for a loan.

The aircraft deal was on the verge of being completed when the bank called: "We don't make aircraft loans." The problem was not a prohibition on aircraft loans; it was the lack of a procedure for an aircraft loan. After discussions with Abbott, a bank officer took a boat loan form and wrote "Piper Cherokee aircraft" in the type of boat block. The loan was approved, and Abbott had his second airplane.

He liked flying the Piper, but his wife liked flying the Piper *to* places. She was the spark that started his search for another airplane. On one long trip she asked, "Can we go faster?"

The best way to go faster was to step up to a retractable-gear airplane. Claude started investigating the possibilities. Mooneys were too cramped. The Piper Arrow was a possibility, but it was 1987 and Piper Aircraft was in financial trouble.

The Bonanza seemed to be the best solution. He'd taken a few hours of dual in one almost 20 years earlier, and still remembered how well it handled. He started investigating the differences between models. He soon settled on the Beech S35,

Fig. 3-7. Claude Abbott, Bonanza owner.

Fig. 3-8. Abbott's Beechcraft S35 Bonanza is the third airplane he's owned.

which was produced in 1964 and 1965. The S model was the fastest Bonanza, with Beech advertising a cruise of 212 mph at 6,000 feet. Because the Bonanza would be used for IFR, he decided to look for well-equipped examples with updated avionics rather than refurbishing a plane with older gear.

The search was on. He read the ads in the aviation newspapers. He flew his Warrior to nearby airports and read the bulletin boards. He finally found a hot prospect at a local field. Bonanza N11JQ was equipped the way he wanted, with full IFR avionics including a flight director, an autopilot, even an oxygen system (Fig. 3-9). But the Beech also sported a tired, run-out engine. Abbott felt the asking price was too high, considering the powerplant ills. He couldn't come to an agreement with the owner, and someone else bought it.

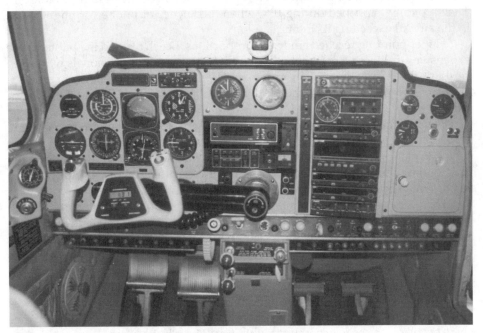

Fig. 3-9. Abbott's Bonanza came equipped with a full IFR panel.

As it turned out, Abbott was right to be concerned with the engine. As the new owner was flying the purchase home, the engine developed problems, forcing a diversion to a nearby field—the field where Abbott's Warrior was based. Abbott ended up trading his Piper for the Bonanza with a broken engine.

The first step was to get the 285-horse Continental overhauled or replaced. Unfortunately, Abbott had used up his ready cash paying off the loan for the Warrior. Back he went to the bank. This time, they loaned him money for an engine rebuild without a murmur. The overhaul cost him about $10,000; while the plane was grounded, he had another $3,000 in work performed.

Four months later, he made his first flight in the rejuvenated Bonanza. His insurance company required at least 10 hours in a retractable-gear aircraft. He already met that requirement, but took several hours of dual to be on the safe side.

Ownership experience

Abbott's ownership of Juliet Quebec has been relatively smooth. Other than the engine rebuild, his biggest maintenance involvement has been the compliance with the tail-strengthening airworthiness directive. A little background on this problem might be instructive. Over the years, the Bonanza's V-tail had been the subject of considerable discussion regarding its strength. On occasion, a Bonanza would suffer in-flight structural failure with the tail separating from the aircraft. The problem was that the failures weren't happening in ordinary flight conditions. They typically happened during IFR, when turbulence or pilot error might result in the aircraft exceeding its limits.

Finger-pointing reached epidemic proportions. Beech blamed the pilots. These expensive airplanes were sometimes bought by moneyed professionals who often couldn't fly frequently enough to remain proficient on a high-performance aircraft. Irreverent bystanders had dubbed the Bonanza the "Forked-Tail Doctor Killer."

Beech gathered examples of each V-tailed Bonanza model and retested them. While they met the certification standard, the margin over that standard was a bit low. The planes were safe when flown at speeds below redline; however, the low drag of the aircraft and slightly lower than nominal pitch stability could lead a pilot into overspeed.

The ultimate solution was a tail strengthening mod, similar to what some aftermarket companies had been offering. In a rare break for normal industry practice, Beechcraft offered to pay for modifying the entire Bonanza fleet. (The cost of complying with ADs usually comes out of the owner's pocket, unless the plane is still covered by the manufacturer's warranty.)

Abbott had known about the problems and the upcoming AD before he bought his Bonanza. After he put 10 hours on the new powerplant, the plane was flown to a Beech-approved facility where the modification was performed. Since then, his plane has been generally trouble-free. The rebuilt engine has provided reliable service. He replaced the hub of the constant-speed propeller during an overhaul; the original hub was rejected due to corrosion. The pitting was minuscule, but in this liability-conscious age, no one wanted to take responsibility for approving it. A used but airworthy replacement cost $975.

Another bite occurred with some minor avionics problems. Abbott turned the plane over to the local Beech dealer. He was shocked when the total bill came to $1,500. Abbott's typical expenses are summarized in Fig. 3-10.

The Bonanza is flown about 50–60 hours per year, mostly local pleasure flights. He and his wife take a couple of long trips each year. "It's got a comfortable 1,000-mile range," he says. "Its speed allows me and my wife to skedaddle from Seattle to the California wine country in 3 hours for a weekend getaway. This would take a day and a half to drive. Our Bonanza is a time machine."

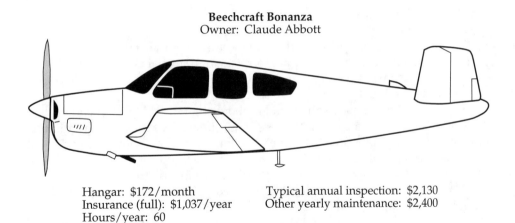

Beechcraft Bonanza
Owner: Claude Abbott

Hangar: $172/month
Insurance (full): $1,037/year
Hours/year: 60

Typical annual inspection: $2,130
Other yearly maintenance: $2,400

Fig. 3-10. The typical yearly expenses of Claude Abbott's Bonanza.

Bonanza N11JQ resides in a closed hangar about 15 minutes' drive from home. Abbott performs very little of the maintenance himself, only replacing bulbs and similar work. He even leaves the oil changes to his mechanic. Although the recommended interval is about equal to the number of hours he flies a year, he has the oil changed every six months. Oil is cheap, compared to the cost of an engine rebuild. Plus, he prefers to have a professional glance over his engine compartment on a regular basis. To him, the $80 charge for the oil change is well worth it.

Because he was careful to buy an aircraft already equipped the way he wanted, he's made few changes. He added a loran and an intercom. He removed a remote-sensing DME (the aircraft already had an ordinary one) and traded it for a couple of good headsets. The aircraft also has suffered a few minor wiring problems, which is typical for an older, well-equipped plane.

Advice

Claude Abbott has a definite philosophy of aircraft ownership. He doesn't scrimp on keeping his expensive high performance aircraft in tip-top shape. For instance, while the Bonanza's all-metal construction lends itself to outdoor storage, he's completely against it. "Keeping a plane outside is a big mistake," he says. He feels the extra cost of a hangar is more than offset by the reduction in maintenance and insurance rates.

"When you look at the overall cost of ownership, gas is nothing." Abbott's Bonanza isn't a 172. Maintenance costs are much higher, as is insurance. The combination of his high fixed costs and moderate annual utilization means he wouldn't see as much of a benefit from autofuel as he might with a less-complex aircraft.

Abbott feels his decision to buy an airplane with a broken engine was right. "The engine is the least important," he explains. "It's better to buy a run-out and get the engine rebuilt *right*. Avionics are a lot more expensive. Pick a plane with the panel that meets your needs. "Understand what you're buying," he adds. He researched the Bonanza series to find the fastest model. "You can't go too fast," he says with a grin.

He pays a premium price for flying the Cadillac of the air. "All replacement parts are based at the going rate for new airplanes," he explains. "If new Pipers sell for $130,000, then parts will be priced in proportion. A new Bonanza will set you back $250,000, and replacement parts reflect that."

Expensive, yes. Some might wonder about the rationality of spending that kind of money on an airplane. "I guess living in a tar paper shack and hiding money under the floor is an option," he says, "but then stories would be told about my eccentricities and bizarre behavior after I'm dead. I'd rather these stories be circulated now. How else can you have so much fun for so little money?"

The Bonanza has met Claude Abbott's needs for seven years, and it looks as though he'll be happy with it well into the future.

NEW AIRCRAFT

One problem with used airplanes: They're used. We can scheme to pick out the best model year. We can bring a prospective purchase in for a thorough prepurchase inspection. We can overhaul the engine, repaint the exterior, and reupholster the inside.

Yet to some people, they're still just old airplanes, probably out of production with the only parts available from aircraft wrecking yards well aware of the aeronautical gold mines they have stacked behind the office.

There's probably not a pilot alive who hasn't wanted a brand-spanking-new airplane. One they can treat right from the beginning. One with a modern panel, equipped from the factory with up-to-date avionics. It doesn't happen very often. Of all the pilots I know, only one has ever owned a brand-new plane, and she won it in a contest (refer to the case study in chapter 7).

The reason, of course, is cost. A brand-new Bonanza costs about $250,000. If I have any friends with that kind of money, they haven't bought me lunch lately. It's not as bad as all that, though. If the Bonanzas are the Cadillac of the general aviation fleet, there are a number of garden-variety Chevy-type airplanes selling for a lot less.

Take Maule, for instance. Occasionally, Maule Air, Inc. has a sale on a stripper model of their four-seat taildragger. Sixty thousand is still a lot of money, but it's not that much more than a 10-year-old Cessna 172 will sell for.

Manufacturer services

One big advantage of buying a new airplane is that you'll be assured of a motivated seller. A private party might decide at the last moment not to sell the airplane. When you do close a deal and write the check, you're on your own.

An aircraft manufacturer, on the other hand, is in the business of selling airplanes. The company will go about this in a variety of ways. Some sell directly from the manufacturer, others through a series of dealers. Being a "dealer" for a particular aircraft might not account for much because the only requirement might be ownership of one example of the aircraft being sold.

One big incentive that the new-airplane maker can offer is a full airframe warranty. For instance, Aviat offers a 12-month/100-hour warranty on the airframe of the Husky and Pitts Special aircraft. Probably nothing major will go wrong, but anything new suffers teething problems.

A warranty gets problems fixed at no cost to you, at least no *eventual* cost to you. A few of the remaining companies have nationwide networks of service centers. Those companies that don't will either have you fly the aircraft back to the factory or will reimburse you for repair work done by a local A&P.

Because the manufacturer covers only the airframe, the engine will be covered by the Lycoming or Continental factory warranty. These companies' warranty claims work similarly to the airframe manufacturers. The avionics situation is similar.

On the subject of money, the manufacturers know the airplanes are expensive. If a finance plan is not available from the manufacturer, they can put you in contact with a number of banks and finance companies.

Aircraft

Years ago, thousands of small aircraft were built every year. Those days and the variety of new planes that spiced up the flight lines are long past. The survivors remain by offering good, although not necessarily modern, products that meet the needs of the market.

That's not to say the selection is completely boring. There are always a number of young companies out to become your light aircraft. Some, like Aerospatiale, are supported by large parent companies. Smaller firms, like Lake, become successful filling a particular niche.

In the few months that this book was written, several aircraft have gained certification. Some of these new companies, like Zenair, have introduced aircraft under the new simplified certification. Aerospatiale's achievement has brought other foreign aircraft into the United States market, trying to emulate the French Aerospace giant's success. The PZL Koliber II is an example, built by a longstanding Polish aircraft company and imported by a stateside distributor.

In the new aircraft arena, the players come and go. As of this writing, Piper has emerged triumphantly from bankruptcy while American General's Tiger production line has been shut down.

Private Pilot magazine publishes a catalog of currently available aircraft in its May issue every year. Table 3-2 is a summary of the 1994 report.

Table 3-2. New production aircraft.

Manufacturer	Model	Engine Make	HP	Gross	Cruise (MPH)	Range (SM)	Seats/Config	Gear	Price
Aerospatiale	TB9 Tampico Club	Lycoming	160	2332	123	506	4	Tri	$113,980
	TB10 Tobago	Lycoming	180	2530	146	751	4 - 5	Tri	$140,980
	TB200 Tobago XL	Lycoming	200	2535	150	745	4 - 5	Tri	$147,170
	TB20 Trinidad	Lycoming	250	3080	189	876	4 - 5	Tri - Retract	$228,300
	TB2T Trinidad TC	Lycoming	250	3080	215	863	4 - 5	Tri - Retract	$264,300
American Champion	7GCGC Citabria Explorer	Lycoming	160	1800	127	575	2 Tan	Conv.	$58,900
	8KCAB Decathlon	Lycoming	150	1800	137	592	2 Tan	Conv.	$68,900
	8KCAB-180 Super Decathlon	Lycoming	180	1800	150	587	2 Tan	Conv.	$72,900
	8GCBC Scout	Lycoming	180	2150	122	783	2 Tan	Conv.	$68,900
American General	AG-5B Tiger	Lycoming	180	2400	164	658	4	Tri	$139,400
Aviat	A-1 Husky	Lycoming	180	1800	140	618	2 Tan	Conv.	$76,495
	S-1T Pitts	Lycoming	200	1150	175	309	1	Conv.	$94,995
	S-2S Pitts	Lycoming	260	1500	175	405	1	Conv.	$109,295
	S-2B Pitts	Lycoming	260	1700	175	319	2 Tan	Conv.	$115,795
Beech	F33A Bonanza	Continental	285	3400	198	822	4	Tri-Retract	$255,800
	A36 Bonanza	Continental	300	3650	202	871	6	Tri-Retract	$311,500
	B36TC Bonanza	Continental	300	3850	224	1132	6	Tri-Retract	$348,700
Bellanca	17-31A Super Viking	Continental	300	3325	205	1205	4	Tri-Retract	$156,500
Classic	Waco YMF Super	Jacobs	275	2950	125	430	2 Tan	Conv.	$196,600
Commander	114B Commander	Lycoming	260	3260	189	725	4	Tri-Retract	$285,000
Lake	LA-250 Renegade	Lycoming	250	3140	152	1093	6	Tri-Retract	$368,380
	LA-270 Turbo Renegade	Lycoming	270	3140	178	1150	4 - 6	Tri-Retract	$404,980
	LA-250 Seafury	Lycoming	250	3140	152	1093	6	Tri-Retract	$398,500
	LA-270 Turbo Seafury	Lycoming	270	2500	178	1150	4 - 6	Tri-Retract	$425,000
Maule	MX-7-180B	Lycoming	180	2500	145	1125	4 - 5	Conv.	$105,840
	MXT-7-180B	Lycoming	180	2500	141	950	4 - 5	Tri	$115,353
	M-7-235B	Lycoming	235	2500	160	859	5	Conv.	$118,062
	MT-7-235	Lycoming	235	2500	159	853	4 - 5	Tri	$135,707
	M-8-235	Lycoming	235	2500	160	859	5	Conv.	$123,369
	MT-8-235	Lycoming	235	2500	160	859	4 - 5	Tri.	$130,369
Mooney	M20J MSE	Lycoming	200	2900	193	999	4	Tri-Retract	$149,725
	M20J ATS	Lycoming	200	2900	193	999	4	Tri-Retract	$167,380
	M20J Ltd.	Lycoming	200	2900	193	999	4	Tri-Retract	$164,900
	M20M TLS	Lycoming	270	3200	256	1208	4	Tri-Retract	$225,750
Piper	PA-18 Super Cub	Lycoming	150	1750	115	460	2 Tan	Conv.	$93,900
	PA-28-161 Warrior	Lycoming	160	2440	145	679	4	Tri.	$128,800
	PA-28-181 Archer	Lycoming	180	2550	148	690	4	Tri.	$135,500
	PA-28-236 Dakota	Lycoming	235	3000	166	828	4	Tri.	$148,800
	PA-32R-301 Saratoga	Lycoming	300	3600	191	937	6	Tri.	$309,800
	PA-46-350 Malibu Mirage	Lycoming	350	4300	259	1139	6	Tri-Retract	$539,075
Quicksilver	GT 500	Rotax	65	1000	81	133	2 Tan	Tri.	$29,995
Star of Phoenix	Phoenix Flyer	Continental	100	1270	120	475	2 SBS	Tri.	$59,995
Taylorcraft	F22 Classic	Lycoming	118	1750	117	639	2 SBS	Conv.	$62,900
	F22A TriClassic	Lycoming	118	1750	121	639	2 SBS	Tri.	$65,900
	F22B STOL 180	Lycoming	180	1750	137	639	2 SBS	Conv.	$69,900
	F22C TriSTOL 180	Lycoming	180	1750	140	639	2 SBS	Tri.	$72,900
Zenith	CH2000	Lycoming	116	1550	115	489	2 SBS	Tri.	$59,990

4

The stuff dreams
are made of
Antiques and classics

THE NONFLYING PUBLIC IS PRETTY FUNNY. To some of them, all little planes are "Piper Cubs," and accidents happen because "the pilot didn't file a flight plan." They're strange in another way, too. Offer to take someone for their first airplane ride, and they'll likely say, "You'll never get *me* up in a little airplane!" Then some add "But I've always wanted to take a ride on an old open-cockpit biplane."

Talk about mixed up! They won't risk their lives in an easy-to-fly tricycle-geared plane with a reliable modern engine; however, they dream about flying in a cranky 60-year-old taildragger powered by an engine built by a company that went bankrupt half a century ago. Still, one can't really blame them. Many pilots would love to fly a fabric-covered biplane from a fresh-mowed field. If some mad fool offered me the choice between an old Stearman or a brand-new Bonanza, I'd be strapping on helmet and goggles before the word "or" left his lips.

Dreaming is one thing. What about reality? That's what this chapter is about. We're going to take a look at some of the joys and difficulties of owning an antique or classic airplane. First off, let's establish what those names signify. Generally speaking, an *antique* is an aircraft built before World War II (Fig. 4-1). A *classic* is an aircraft built between the end of the war and up to about 1955. Let's take a look at antiques first.

ANTIQUES

What's it like to own an antique airplane? Probably the best analogy I can make is to old cars (Fig. 4-2). While I was growing up, my father talked about buying a Ford Model T just like he had when he was young. Finally, when all the kids were out of college, he picked up a Model A and a '31 Chevrolet. But he never forgot the tin lizzie. He scoffed at modern reproductions: "The pedals are wrong." "It's just got a Pinto engine in it." "Look at those modern wheels and tires."

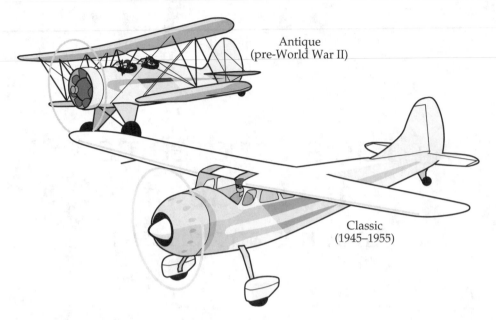

Antique
(pre-World War II)

Classic
(1945–1955)

Fig. 4-1. Antique and classic aircraft are two different categories.

Fig. 4-2. Owning an old airplane is much like owning an old car. This Beech Staggerwing and Cord automobile are contemporaries.

Finally, he bought a '23 Model T. All the running gear had been brought back to original condition, with a spiffy two-tone paint job replacing the original "Any color you want, as long as it's black" scheme. Just what he'd wanted all those years. As the saying goes, "Be careful what you wish for . . . you might get it."

In the light of 50 years of progress, the Model T is a pretty lame automobile. The brakes are mechanically operated, which is far less effective than the hydraulic brakes of later cars. It worked fine on dirt roads in 1935, but driving the Model T is hard work in today's stop-and-go traffic. Its suspension was designed to be robust and sturdy, not smooth. The T rode like a paint-shaker on steroids.

Starting problems had been relegated to fond memories. But now he had a *real* Model T, with fuel and spark controls, and a manual crank eager to break one more arm before being relegated to the scrap heap.

Because it was so uncomfortable to drive and ride in, Dad bought a trailer and towed it to parades and car shows. On nice summer days, he cruised the neighborhood with the top down. Parts were no problem because there are a number of companies supporting the antique market. In any case, it didn't spend much time on the road. Dad had fun; he enjoyed the way it drew attention and the jaunty way it traveled down the (preferably smooth) street.

But he sold the Model T after two years and replaced it with a 1953 Ford Crown Victoria Coupe. A car with power brakes. A car that rode like a pillow in a snowfall. A car with a starter, and a *heater*. One he could drive to work and show off to his buddies, or take on long trips.

Antique airplane ownership really isn't much different.

Handling

Ninety-nine percent of your antiques are going to be taildraggers, and it'll be conventional gear at its worst. These planes were designed when all it took to be an aircraft designer was a piece of chalk and a hangar floor. Nowadays, a designer knows how to minimize a taildragger's quirks. Those old biplanes were designed with tail skids and grass strips in mind. Putting a wheel on the back end doesn't make them civilized.

By now, most but not all of the regularly flown antiques have brakes. Often, though, they don't work very well. Brakes that are too effective can contribute to noseovers, so they're sometimes deliberately kept soft. You might not have enough brake power to stop if something suddenly appears in front of you.

But that's kind of moot in a lot of the old biplanes because you won't see it anyway. As illustrated in Fig. 4-3, forward visibility is often nonexistent while on the ground. The pilot usually sits far back from the nose; on takeoff, he can't see the runway until the elevators become effective enough to raise the tail. Pilots of old biplanes often fly a curved final to allow them to see the runway until the last possible second.

Visibility doesn't get much better in level flight. Something blocks the view in almost every direction on a biplane, which combines the visibility drawbacks of low- and high-wing configurations. The pilot has to keep his or her head bobbing to see around the obstructions.

Fig. 4-3. Periscope up! While *Samson* isn't an antique, it aptly demonstrates the terrible forward visibility from the pilot's seat of many old radial-engined biplanes.

The pilot is working hard in other ways, too. Some of these old planes' handling characteristics are throwbacks to the bad old days. You're probably familiar with adverse yaw, where the "downgoing" aileron generates more drag than the "upgoing" one. The effect makes the plane tend to yaw in the opposite direction of the turn.

Modern airplanes compensate for it. Cessnas have Frise ailerons, which increase the drag caused by the downgoing aileron. Others gear the upgoing and downgoing ailerons differently to eliminate adverse yaw.

But take the famous Jenny, the Curtiss JN-4D. Designers hadn't really known about the effect when the big biplane was developed back in 1913. Pilot reports say that when the stick is moved left, adverse yaw is so bad that the *right wing drops*. Pilots must lead the turns with massive amounts of rudder.

Other flight characteristics? Well, the stall probably won't be of the gentle Piper variety. The inventor of the stall warning horn probably was still in short pants when antique planes were built. So don't expect much warning. Many of these older planes will be quite happy to spin, too.

Parts and maintenance

One of the biggest impacts on the maintenance expense for antique aircraft is their out-of-production engines. Sixty years ago, the words Menasco, Ranger, OX-5,

Kinner, Warner, and Hispano could be found on cowlings all across the country. Even garden-variety parts like spark plugs have become rare. Your ability to keep these old planes running (engines and airframes) will depend upon a number of factors besides money.

How rare an antique is it? Most of the spare parts for antiques are removed from junked or crashed planes; therefore, if the plane had limited production, your chance of finding a wreck for parts is greatly reduced. Like the Ford Model T, if an airplane type is popular and many were made, parts might not be that difficult to find. The same holds true for the engines themselves; Kinner engines are popular in a number of antiques. Even if you have a rarer bird, if it mounts a Kinner, you've got a better chance of keeping it running.

The question is, is it worth keeping running? Nostalgia tends to hide the fact that some of these old aircraft engines weren't worth a darn when they were new. Some of those early magnetos, for instance, have extremely poor reliability. The FAA's requirement that aircraft engines have two ignition sources can be traced to this era.

A classic example of the questionable reliability of older engines is the Szekely three-cylinder radial, seen in Fig. 4-4. Producing 35 horsepower, it was used in a number of 1930s planes, including the Curtiss Junior, the Buhl Pup, and even a model of the Piper Cub. If you see one at a fly-in, you'll notice cables or rods circling the engine atop the cylinders. An AD required this modification because the Szekely had a bad habit of cracking at the cylinder bases and *throwing the cylinders clear of the engine*. The cable doesn't stop the cracking, it just holds the engine together so the pilot can nurse the plane down to a safe landing!

Fig. 4-4. The Curtis Junior was powered by the trouble-prone Szekely radial. White rods around and between the cylinders keep the engine from flying apart if a cylinder breaks!

Even if your antique has a better-designed engine, you can't expect these old planes to be as reliable as their modern counterparts. Their TBOs are likely to be a quarter of that of a modern Lycoming. They'll also require more work on your part; some require greasing the rocker boxes before every flight, for instance.

Owners' groups are good sources for parts and information; they often set up an excellent network. Few clubs require that members actually own the aircraft. Join the group before you buy; get involved with their activities. Even reticent owners tend to get more expansive when around fellow enthusiasts. Listen and learn. As Yogi Berra said, "You can hear a lot, just by listening."

Other factors

You'll spend a lot to buy your typical antique; you'll undoubtedly want to insure yourself against loss. Factor the higher premiums into your ownership equation.

For that matter, you wouldn't dare keep an antique plane tied down outside or in an open hangar. Expect to pay more for safe storage. To make matters worse, many antiques are big airplanes. The ordinary general aviation hangars available in your area might not be large enough.

Plus, moving these big airplanes around is a major undertaking, one made even worse if the hangar is a tight fit. You might have difficulty handling it on your own (Fig. 4-5). A winch might help. On the plus side, though, you'll never be lacking for willing passengers who can help you drag the plane around.

Fig. 4-5. Big antique biplanes are hard to handle alone. This Stearman just barely fits inside an ordinary general aviation hangar.

The good news

Not all the news about antiques is bad. By the mid-1930s, aircraft designers were getting the hang of it. Conventional handling is more and more common to airplanes built in this period. Not all the planes are large and bulky. As mentioned earlier, the Piper Cub was in production in this era.

In many cases, modern engines have been substituted for unreliable originals. One Buhl Pup that was originally equipped with the Szekely engine is flying with a Continental A-65, which is the same engine mounted on the Aeronca Champ.

The parts problems for antiques aren't unsolvable. Most of these planes are so simple, built using basic techniques and low-tech solutions. If a rare part is needed, it can usually be made. Many antiques contain a surprisingly low percentage of truly original parts, especially wood components. The use of modern epoxies and metal alloys also serves to increase the strength and reliability of these old birds.

And parts, especially parts for certain engines, might not be as rare as you might think. Far-sighted people started stockpiling parts years ago. Some vendors have started limited manufacture of certain crucial items. Plus, several engines from the immediate prewar era had massive production runs after Pearl Harbor. Not just for airplanes, either. For instance, the 220 Continental radial powered some World War II tanks! True, after 50 years, these sources have started to dry up. But an overhaul of the Continental radial still costs about the same as a Bonanza engine.

While it's true that some antiques have to be babied and should only be flown occasionally, others are less delicate. A Stinson Reliant or Fairchild 24 is just as good as a Cessna 172 because in most cases you can preflight, load them up with a couple of friends and baggage, and take a long trip.

Antique summary

You didn't expect it would be easy, did you? Or even cheap? Still, there's a lot of satisfaction in owning an antique. There's camaraderie with fellow owners. There's the plain ego value of watching the parade of admirers at fly-ins.

The best suggestion, if you're interested in antique ownership, is to thoroughly research the history and availability before buying. There are a number of books detailing the design and production of certain historic aircraft. Join the clubs well before putting your money down. Talk to owners; find out about parts availability and reliability. Get known within the circles; find out who's ready to sell before they take out advertisements.

Getting known within the circle can make a big difference. Some antique owners are, well, a little possessive about their airplanes. They care about what happens to the plane, even after they no longer own it. If you can establish yourself as a responsible person, you might get a good deal from someone looking for a good home for his or her pet plane. I know a case where an owner refused to sell his plane to a certain prospective buyer. The prospective buyer, it seemed, had a tape deck installed in his current airplane. The seller felt that anyone who listened to music while flying wasn't paying enough attention. He didn't want *his* airplane damaged by some sloppy new owner!

It might seem a strange attitude to those who have never owned an airplane, but to those who have flown or owned older airplanes, modern Cessnas and Pipers are somewhat sterile. They can't match the personality of the old-time machines.

Antique aircraft, like antique cars, have a feel all their own. If one can afford the price of admission, the cost of a flying antique is well worth it.

CASE STUDY:
PARTNERS IN A CLASSIC BIPLANE

Little did Don Connell and Stan Brown (Fig. 4-6) realize their interest in radio-controlled (RC) models would lead to ownership of a full-scale Boeing Stearman PT-17. They weren't just Sunday RC fliers, though. They were wrapped in the competitive world of RC pylon racing. One side benefit: The races were often held on small airports. Both licensed pilots, they got into the habit of wandering around the tiedowns and hangars after the meets were done.

Fig. 4-6. Stan Brown (left) and Don Connell.

One day in 1986, Stan called Don, "Wanna buy a Champ?"; $7,200 later, they were airplane owners. The RC flying was put on the back burner. Connell says, "Once we were in full-scale, neither of us had time for two hobbies."

The Champ was nice, but the owners got the bug for a bigger airplane. "We started looking for Tiger Moths and Chipmunks," recalls Brown. Then they flew the Champ to the annual Stearman Fly-In at Cottage Grove, Oregon. They listened to the growl of old engines all weekend. They talked to owners and discovered that parts weren't much of a problem for the big Boeing biplane. Stearmans got into their blood. Brown said to Connell, "Maybe we're crazy if we *don't* buy a Stearman!"

They set themselves a $40,000 limit and grabbed a *Trade-A-Plane*. "We called every ad and got pictures," says Stan Brown. "I went to look at one good prospect in Texas, but the owner sold it just before I got there." They started despairing of finding a Stearman in their price range.

They finally found one in Florida, a full continent away from their homes in Seattle. They hopped an airliner to Florida and looked the plane over. It had been recently restored. Its owner had multiple aircraft and was selling the Stearman just to clear some space in his hangar.

Connell and Brown bought the big biplane (Fig. 4-7) for $37,500. They then flew home via airline and called Bill Sato, a good friend with lots of Stearman time. Sato accompanied Connell to Florida and checked him out on the airplane. The two then flew the plane to Denver, where Brown met them. Sato gave Brown dual instruction the rest of the way home.

Fig. 4-7. Connell and Brown bought their 1942 Stearman in Florida and flew it home to Washington state.

By their triumphant arrival in Seattle, both met the insurance-company dual requirement. They looked forward to a late-fall of open-cockpit fun, but the landing gear failed three weeks later. Fortunately, it was not a complete failure, but 40 years of rust had critically crippled the left landing gear leg. "Any bad landing coming across the United States could have folded the gear," recalls Brown.

Connell just shrugs: "That's what happens when you fly old airplanes." Small consolation. They borrowed a fork lift to hoist the plane to remove the landing gear. They set it atop barrels while the gear was rebuilt. The Stearman looked like a big orange Bantam brooding rusty eggs. One passerby commented: "It'll never fly, it's just an old ramp rooster."

Six weeks and $6,000 later, Stearman N1035M was ready. In addition to the newly rebuilt landing gear, the biplane's nose and tail were graced with red, white, and blue trim. In addition, the fuselage now sported a belligerent cartoon chicken and a brand new name: *Ramp Rooster* (Fig. 4-8).

Ownership experience

Ramp Rooster's owners have a "handshake" partnership agreement on the airplanes, where both have agreed to split the costs. Their long-standing joint ownership of the Champ let both men know they could trust the other. "The AOPA partnership agreement looks good," says Connell. "I think if I went partners with anyone but Stan, I'd use it."

Beyond the initial problems that were taken care of, the Stearman has been relatively trouble-free. "Hardly anyone wants to annual a Stearman," says Connell. "But it's getting to the point where hardly any IA wants to tackle *any* fabric airplane, including our Champ."

When they find a willing IA, though, the process is pretty straightforward. Stearmans are very simple machines, with parts sized for military wear and tear. The 120 hours that *Ramp Rooster* flies each year hardly shows. Connell and Brown's annuals cost $200 or so, with a ride for the IA thrown in. Note that this is much lower than even a typical Cessna 150 annual runs, which is an example of the advantages of the informal "old boy" network that has sprung up in the antique aircraft movement.

All other maintenance is performed by the two owners, under the supervision and signoff of a local A&P. Their maintenance costs in the typical year run about $1,000. Unfortunately, 1993 was not a typical year. *Ramp Rooster's* 220 Continental radial lost a rear main bearing. The partners started searching for a new engine.

Luck came their way. They found a Fairchild owner who'd had a custom overhaul performed on a 220, but couldn't get an STC to install it on his airplane. The partners were able to pick up this better-than-new engine for $10,000. A new wooden Sensenich prop completed the effort, and they were back in business.

They keep *Ramp Rooster* in an open hangar, paying a rent of $122 per month. Insurance (including hull coverage) is about $1,100 per year through the Experimental Aircraft Association.

For all its size and heft, the Stearman isn't as much of a gas hog as one might assume. The Continental burns 12.5 gallons per hour: "You can set your clock by this thing," says Connell. They have an autofuel STC.

In no sense can *Ramp Rooster* be considered a working machine. The majority of its 120 hours a year are local pleasure flights, although the owners have flown

Fig. 4-8. Col. Sanders wouldn't have anything to do with this tough old bird. No Stearman has ever suffered in-flight structural failure.

to the annual Stearman rally in Galesburg, Illinois. *Ramp Rooster* makes an excellent air-to-air photo platform (to which this author can attest). It was also featured skimming over the eastern Washington prairie in a video called "America by Air."

Ramp Rooster sometimes gets its way paid to local fly-ins. Some organizers are more willing to pay gas costs for "civilian" airplanes than those with military paint jobs. But the Stearman isn't a gussied-up show bird that leaves the hangar only when there are awards to be won. It growls aloft on any nice evening or sunny weekend. Connell and Brown treat it like an *airplane*, not a trophy magnet. A summary of their typical costs is shown in Fig. 4-9.

Boeing Stearman PT-17
Owners: Don Connell and Stan Brown

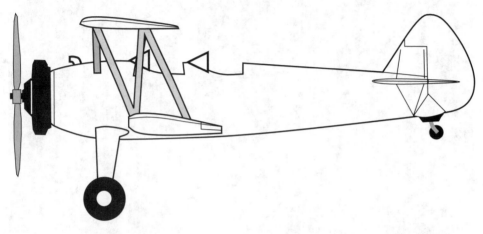

Hangar: $122/month   Typical annual inspection: $200
Insurance (full): $1,105/year   Other yearly maintenance: $1,000
Hours/year: 120

Fig. 4-9. Typical ownership costs for *Ramp Rooster*.

Advice

Stan Brown and Don Connell don't fit the mold of "gentlemen" antique aircraft operators. They don't keep it in a temperature-controlled hangar on their private field. Brown is employed, and Connell is a retired engineer.

They keep *Ramp Rooster* flying through infusions of sweat, not bundles of cash, and by developing contacts that help them pare the Stearman's bite on their pocketbooks. "Some guys just come out to the airport just to fly," says Brown. "When inspections come around, they take their planes to the high-priced FBO down the street." Connell and Brown are deeply involved with the airport's social circuit. "When a guy goes out to buy an airplane," says Brown, "he becomes part of the airplane 'group.'" And you can use that group to cut your cost of ownership.

For instance, their contacts found a place that sold retreaded tires for their plane, at an 80-percent savings. Repairs of one minor problem were initially quoted at almost $1,000, until a friend suggested an $80 alternative. You don't make those contacts while renting a 172 twice a month!

"Don't wait until you're 40 to start flying," suggests Connell. "Sell whatever you have to so you can buy an airplane." Brown agrees. "Find something cheap to fly, just so you get accepted by the flying crowd. The more you love airplanes, the earlier you should get into it."

CLASSICS

Recall that classic aircraft cover the period from the end of World War II to 1955. A new category has come into prominence recently: "contemporary classic." This category acknowledges that there were a number of aircraft in the late 1950s that were still holdovers from the old days; even the early 172s look quite different from the later models. The current cutoff year is 1961.

Classics are a nice way to fly. Most of them are powered by modern aircraft engines, Continentals and Lycomings that any mechanic can maintain and find parts for replacement. Many airframes are all-metal, which precludes problems with fabric deterioration or wood rot.

Drawbacks are few, compared to antiques. As with any older piece of machinery, parts keep getting rarer. Good new parts, though, are being produced by several companies. Univair Aircraft Corporation in Aurora, Colorado, owns a number of old-aircraft type certificates. You can practically build a Stinson or Luscombe from their parts catalog.

With Wag-Aero in Lyons, Wisconsin, you *can* build a Piper Cub. Wag-Aero's "Sport Trainer" kit is a Cub replica; all of the major parts are the same as the real Piper. They also sell approved replacement parts for much of the Piper J- and PA-series.

The major drawback is that classic aircraft are more likely to be in rather sad mechanical shape. They've traditionally been the choice for those flying on a shoestring. As such, they've been outside for 20 years and received only enough maintenance to keep them airworthy. If that.

There's a silver lining, though. First, these are the lowest-cost planes out there. You can't really pick them up for a song anymore, but it doesn't take an entire choir. Second, the refurbishment and restoration of these older classics is becoming more and more common. A lot of mechanics and other specialists are quite familiar with the innards of these older birds.

All it takes is money, of course, but perhaps not as much as you might think. Most of the work involved in restoring these planes is grunt work that almost anyone can do. Plenty of folks have restored their own aircraft. By working with a local A&P, you can perform almost all of the work involved and end up with a show winner. This next case study is a good example.

CASE STUDY: FROM RAUNCHY TO RAVISHING

When you bought this book, you probably had the fond dream of finding some *really cheap* airplane that you could buy and fly. The thought struck Mike Furlong

Fig. 4-10. Mike Furlong is a design engineer with a major aerospace company.

(Fig. 4-10) one Sunday morning while reading the classified ads in his hometown newspaper. There was a Stinson 108-3 Station Wagon listed.

Mike didn't know much about Stinsons. He hadn't thought about buying a plane at all. But the ad made him realize how tired he was of renting Cessnas. It was time to buy his own plane. The best thing was the price: $3,500! (Cheap? I bought my Cessna 150 at about the same time. I paid $6,000 and thought it was a good deal.)

Furlong called the Stinson's owner. The plane hadn't flown for a year and a half. "Haven't been able to afford it," sighed the man. "Flew just fine, though." Furlong drove 60 miles to the plane's base. Stinson N4095C looked a bit ratty, but then it had been sitting without moving for almost 2 years. He bought it.

With a little bit of work, he got the engine running. He knew better than to fly it. It didn't have a current annual, and he didn't have any taildragger time. Still, it shouldn't take much. A little clean-up, an annual, and a taildragger checkout. Nine years and $20,000 later, Mike flew the Stinson for the first time.

The resurrection of Charlie

Mike Furlong got his first inkling of trouble when he took *Charlie*—nicknamed from the last letter of the N-number—to a local A&P for the annual and the man just laughed. The fabric was shot.

Mike wasn't too downhearted. He'd known the covering was marginal; he'd just hoped to get a year or two of flying out of *Charlie* first. He brought a trailer and started to disassemble the plane. It took 2 hours to remove one wing's anchor bolts, which had corroded in place.

Mike eventually got the plane home. The farther he looked into *Charlie's* innards, the worse it looked. The Stinson would require a complete rebuild. He didn't really have enough room to do the job, but his father, a long-time aviation enthusiast, became intrigued with the project. Bill Furlong built a combination shop/garage on his property, and Mike contemplated the long struggle ahead of him.

So far, he'd done just about everything wrong. He hadn't researched Stinsons. He saw the ad, liked the price, and bought the plane. "I just got a wild hair," as he describes it. A simple prepurchase inspection would have saved him years of work and aggravation.

But from the day that the resurrection of *Charlie* began, Mike did everything right. It started with the shop. He and his father ensured that it was painted and well-lit, big enough not only for the airplane but for safe, dry storage of the removed parts. Too many people have tried to restore airplanes in single-car garages; the Furlongs went about it the right way.

Mike then bought a ton of film for his Polaroid camera. He didn't know beans about Stinsons, and when *Charlie* was disassembled into its smallest components, Mike knew he'd be hard-pressed to remember how to put it back together. He shot pictures as he took it apart: dozens of pictures. He'd scribble on the print with a pen, highlighting the location of other components to be sure he'd know exactly how to reassemble it when the time came.

Sure, there are manuals, but they cover the big things, not how some little piece of trim might lie on the cockpit sidewall. When the time came, Mike wanted to put the plane together right.

"Organization was important," he says, with considerable understatement. He didn't just throw the parts in the corner when they came off. He bought 300 zip-lock plastic bags, and stored his parts in them. All the screws for a certain panel would go into a single bag, along with a tag indicating exactly where they'd come from. Even though few would be reused, Furlong wanted to be able to tell *exactly* what kind of screw to buy. He filled two large cardboard boxes with plastic bags full of fasteners.

The repairs began. The rotted wood was replaced. The rusted-out steel tubing in the aft part of the fuselage was cut out and new 4130 welded in. Old parts were cleaned and reused; new ones were found if necessary.

He eventually transferred the plane to his own house. The fuselage and tail went into the garage, but there wasn't adequate room for the wings. Only one place in the house was big enough to take the 15-foot panels: The family room. *Charlie's* wings stayed there for 4 years. Fortunately, his wife supported the project. "The wings kind of leaned in the corner, tucked next to the wall," says Mike.

But he couldn't maintain the pace of restoration. Things started to slow. He was getting to the point where he needed expert advice, and he couldn't really afford to hire an A&P.

To top it off, he'd been having problems with his boat. He put it up for sale. A prospective buyer, an older man, came in response to the ad. He glanced into the garage. "Stinson 108, huh?"

Turned out the visitor was a retired A&P with an inspection authorization (IA). He offered an even swap: the boat in exchange for supervising the rest of the way on the restoration, and he'd provide the crucial signoff.

Charlie's rebirth entered the final leg. Furlong shipped the Stinson's 165-hp Franklin engine to a rebuilder specializing in these rare powerplants. He covered the aircraft with Stits fabric and painted it dark maroon with cream trim. He sent the seat covers and the interior panels to a local upholstery shop. Rather than spend time and money working on avionics, he junked the antiquated navcom and installed a simple handheld radio.

Mike towed the fuselage to the airport on its wheels. The wings followed on a trailer. Mike, his father, and the IA reassembled the Stinson.

Finally, the plane was ready. Furlong flew *Charlie* (Fig. 4-11) for the first time in May 1993 with an instructor in the right seat. After 7 hours of dual, he was turned loose on his own. Mike and *Charlie* flew more than 80 hours in the first 8 months.

Fig. 4-11. It took more than $20,000 to get *Charlie* in flying condition.

Ownership experience

Stinson called the 108-3 a Station Wagon for good reason. It's a true 4-seater, with a cavernous baggage compartment and a removable rear seat. For a total expenditure of 9 years and more than $24,000, Mike Furlong ended up with a practically new Stinson 108-3 that flies faster and carries 200 pounds more than a Cessna 172.

Like most new machines, it started out with a few bugs. Oil leaks are typical in newly restored airplanes, but that doesn't make dirty streaks on new paint any easier to take. There are always some fittings that aren't quite snugged up—in Mike's case, the valve covers.

Stinson 108-3 Station Wagon
Owner: Mike Furlong

Hangar: $60/month
Insurance (full): $1,350/year
Hours/year: 80

Typical annual inspection: $500
Other yearly maintenance: $200

Aircraft fully restored by owner over nine-year period

Project cost summary

Description	Cost
Initial purchase	$3,500
IA hire	$2,215 (includes value of boat and motor traded for services)
Engine & accessory rebuild	$7,220
Hardware	$2,765
Fabric, finish, and wood	$4,740
Avionics and instruments	$2,630
Tools (not including welder, welding supplies, or air compressor)	$1,160
Total:	$24,230

Fig. 4-12. A summary of Furlong's ownership expenses.

In addition, *Charlie's* Franklin engine has the oil-filled crankshaft necessary for a constant-speed propeller. The crankshaft was plugged because Furlong installed a new fixed-pitch prop. The plug leaked, and because the plug was right behind the prop, oil was flung everywhere. Mike eventually got the leaks stopped.

He runs 80-octane fuel in his Stinson. His engine rebuilder advised against using autogas in Franklins because some valve problems have been caused by autogas. Parts are rare for the 165-hp engine, so Furlong intends to follow the expert's instructions.

Mike's quite happy with the aircraft's performance and value, even if it did take a lot of work. "The Stinson was half the price of a used Super Cub," he explains. It gets almost the same performance as Piper's workhorse, but with a higher useful load.

The aircraft resides on an island airport that is a ferry-ride away from Mike's home. The island hangar costs $60 a month; he's been on the waiting list for the local airport for almost 2 years.

One financial item gave him an unusual surprise: "The insurance was lower than I expected." Liability insurance and $20,000 of hull coverage costs Furlong $1,350 a year. The cost will be going down for the second year of coverage because Mike will have plenty of taildragger hours at renewal time.

He did go through some convolutions to get coverage, though. The first company he called required that he have 15 solo hours in the airplane before they'd cover him. The second required just 5 hours of dual and 5 solo hours.

Furlong anticipates minimizing the cost of annual inspections by participating in the process. After all, who knows the plane better? Figure 4-12 summarizes Niner-Five Charlie's resurrection and ownership.

Advice

Mike knew *Charlie* had problems when he bought the Stinson. He expected some amount of work would be necessary because the plane had not been flying very recently, but had flown as recently as 18 months earlier.

He ended up with a gorgeous Stinson, but if he had known the plane's true condition, he would have passed it by. He was too enthusiastic about the prospect of owning his own plane.

"Sleep on it," is Mike Furlong's simple advice to the prospective aircraft buyer. "Get a prepurchase inspection."

5

Roll your own
Homebuilts

THE GENERAL AVIATION INDUSTRY has been in a slump for more than 10 years. Deliveries of new aircraft have dwindled. Cessna stopped making small planes. Piper declared bankruptcy. American factories produced few, if any, totally new aircraft.

However, that doesn't mean we haven't seen a bunch of hot new planes. Interest in homebuilt aircraft has grown from a flickering campfire to a roaring inferno of plans that range from low-and-slow fun airplanes to 300-mph fiberglass speedsters (Fig. 5-1). The Experimental Aircraft Association's annual fly-in convention at Oshkosh is the world's largest aviation show. Every year, new homebuilt designs vie for the attention of a rising number of people intent on building their own plane.

How big is the homebuilt phenomenon? In 1991, three times as many light aircraft were built in garages than on factory floors. A number of homebuilt companies have sold more than 2,000 kits each, kits that cost tens of thousands of dollars each, not just plans that might sell for several hundred. But there's more to a homebuilt than signing a check and spending a couple of weekends in the garage.

HOMEBUILT BASICS

To begin, let's establish the exact definition of a homebuilt. Back when you were taking flying lessons, your instructor showed you a little plastic pouch in the back of the airplane. Inside were a couple of slips of paper. "You gotta have these on board to be legal," your instructor said. "Look, here's the registration, radio license, and the airworthiness certificate." The instructor then stuffed the forms back in the packet. You probably never looked at them again.

Two of them cover mere bureaucratic formalities. The registration verifies that the FAA knows about that particular airplane, and the station license authorizes a pilot to use the communication radios. The airworthiness certificate has a story of its own.

Fig. 5-1. In 25 years, homebuilts went from wood-and-fabric sport machines to fiberglass speedsters.

The factory that built your trainer spent thousands of manhours to be able to put that little scrap of paper in your airplane. To earn certification in the *normal*, *utility*, or *aerobatic* categories, the manufacturer proved the plane could withstand certain G-forces and possessed certain safety and warning equipment. Its flight characteristics had to meet stringent government requirements. Its engine came from the narrow ranks of FAA-approved powerplants. The government sent its own test pilots to verify the findings of the company.

The reward came when the FAA awarded the design a *type certificate* (TC). The airworthiness certificate crammed into that flimsy holder guarantees that this one particular plane met all the requirements of the TC.

The next time you're around that little plastic packet, take out the certificate. Smooth the wrinkles; unfold the dog-eared corners. Many of our fellow pilots and dozens of my fellow engineers worked hard to put that tiny piece of paper in your plane.

The experimental category

All that effort is expensive. It can only economically be put into production aircraft, where the cost can be amortized over hundreds of planes. The FAA recognizes that certain limited-purpose airplanes don't really require such extensive analysis and testing. For these airplanes, the FAA provides the *experimental* category. Aircraft with Experimental certificates don't have to prove anything before they're allowed to fly. The FAA only demands that precautions be taken to protect the lives and property of those not operating the aircraft.

The experimental category includes a number of subcategories, including *racing*, *research and development*, and *amateur-built*. Of these, amateur-built has the most relaxed rules. Aircraft in this category are restricted against flight over congested areas (except to and from airports), flight for hire, or night/IFR flight unless specifically approved. To qualify for the amateur-built category, at least 51 percent of the labor required to build the plane must be done by a nonprofessional builder "for educational or recreational purposes."

Homebuilt advantages

Aircraft design, production or homebuilt, is a study in compromise. The designer merges the performance objectives, FAA rules, customer preference, and marketing inputs to generate an aircraft design. She cannot hope to please everybody; rather, the designer's goal is an aircraft that most can accept with minimal grumbling.

But experimentals aren't bound by the FAA's conservative design requirements. The design can be optimized toward any specific goal that the designer wants. With the ability to optimize the aircraft toward a particular goal, the homebuilt-aircraft designer can produce aircraft that outperform production planes (Fig. 5-2).

Fig. 5-2. Jim Smith's Sonerai IIL out races Cessna 150s on the power of its converted Volkswagen engine. He built it for less than $8,000.

Speed is a typical homebuilt design goal. Take the Lycoming O-235 engine mounted in the Cessna 152. Several homebuilt designs use this engine; one design cruises *100 mph faster than the 152!*

There's no magic involved. That particular homebuilt has one quarter less wing area, retractable gear, and the occupants sit semireclined. The aircraft's drag is but a fraction of the Cessna trainer's.

Homebuilts cost less to operate, as well. If you owned a 152, you'd be spending money every year for a mechanic to perform the annual inspection, and if something broke, you'd have to replace it with parts approved for aircraft use.

Not so with a homebuilt. If you built it—at least 51 percent—the FAA will give you a *repairman certificate*. This authorizes you to perform all the maintenance and inspections on your bird. You can replace a broken component with anything that you feel is safe. You can even use a converted car engine and avoid excessive overhaul costs.

Acquisition costs are cheaper than those of a new production bird. A Glasair III might cost $50,000 by the time you're finished, but its performance exceeds that of a $200,000 Bonanza. Plus, you could buy the components of that kit gradually over time, stretching out the cost over a number of years. This has the effect of a no-interest loan, sort of a layaway/flyaway plan.

Finally, production aircraft are, well, rather boring. They're the equivalent of four-door Chevy sedans. A Cessna 172 or Piper Warrior might be good transportation, but they do not cause very much unbridled excitement.

Homebuilts are the sports cars of the air. Flight controls are light, and they're fun to slice through the sky. There's a lot of satisfaction involved in building your own plane, and you'll be sure to draw a lot of attention.

Homebuilt disadvantages

When compared to new factory aircraft, the cost of building a homebuilt looks like a real good deal; however, few people actually buy new production planes. The homebuilt's biggest competitors are used certified aircraft, and the homebuilt often comes out second best.

You can build an IFR Glasair II or Lancair 320—both are composite-construction speedsters—for $40,000–$50,000 and cruise about 200 mph. For the same money, you could buy an early 1970s Mooney and carry two more people only a bit slower—tomorrow, not 5 years from now.

It's the same for aircraft on the other end of the speed scale. The $18,000 or so you'll spend to build an Avid Flyer or Skystar Kitfox could pick up a nice Champ.

And make no mistake about it: It doesn't matter if the plane comes as a kit; building an airplane is a lot of work. Don't be misled by the construction estimates from the kitplane companies because in the real world, builders often double or triple the kit manufacturer's advertised construction time.

You'll also need somewhere to build the plane. Homebuilts have been constructed in apartments, but it's certainly not the optimal course. There'll be tools to buy, too.

Another worry: When your plane is registered, you are the legal manufacturer. If you sell the plane and the buyer, or any subsequent buyer, crashes it, you might be sued.

Finally, homebuilts have come under a bit of controversy regarding their handling characteristics. Remember, aircraft design is the art of compromise. Designers of production aircraft usually sacrifice performance in the interest of meeting the FAA's handling and stability requirements. There is no free ride in aircraft design. If you push one side of the envelope out, another side squeezes in.

Take our humble O-235-powered Cessna 152. The same engine was installed in an earlier production aircraft, the American Yankee. The Yankee cruised 10 mph faster and had crisp, sporty handling. It also has one of the highest accident rates of the general aviation fleet. Its fatality rate is *three times* that of the Cessna 150. The Yankee has been particularly faulted for its poor stall characteristics.

So what happens when you push the speed further? The best illustration comes from Australia, where homebuilt designs must be flight tested and approved. One of the United State's most popular homebuilts flunked, and it was not flunked in a seldom-seen corner of the flight envelope. When an experienced Australian test pilot stalled the aircraft, it flicked inverted. A thousand feet were lost in the recovery with a *professional test pilot* at the controls. Several other popular United States designs have just barely passed muster down under.

Don't be frightened away. This means that homebuilts might fly differently than the production planes you're used to. They'll be a little more challenging to fly precisely, but most of them will be safe. Take that American Yankee, for instance: a death trap? Not if you take a few hours of lessons from an instructor well-versed in the Yankee's idiosyncrasies. I did, and found the Yankee a joy to fly. For fun flying, I'd pick the Yankee or one of its Grumman descendants over a 152. Interestingly enough, the Yankee was originally designed as a homebuilt.

Should you build?

Don't be too concerned about safety issues. Homebuilt-aircraft accident rates are similar to those of production aircraft. Structural failure is almost invariably caused by deviations from the plans or exceeding the aircraft's flight envelope.

As far as stability and handling, research the aircraft before you pick a kit, and get a good checkout before your first flight. Homebuilts aren't difficult, only different, but it still takes a lot of effort to get to that first takeoff.

The question is, do you want to *build a homebuilt*, not just own one, not just fly one, but do you actually want to go through the years of effort (Fig. 5-3) required to construct your own airplane? If you're looking at the construction process merely as an intermediate step to flight, you'll probably fail. If you want to fly a plane, *buy a plane*. If you want to *build a plane*, then build one. It's as simple as that.

If you really like the designs, but decide that you really don't want to build one, take heart because subsequent sections of this chapter discuss buying used and partially-completed homebuilts.

Fig. 5-3. Buying an airplane takes just a day; building an airplane takes years. Are you sure you want to wait?

HOMEBUILT CONSTRUCTION

Can you build your own airplane? Probably. In the early days, homebuilders mostly came from the ranks of those with considerable shop skills. Folks who were willing to spend thousands of hours building a simple single-seat sportplane.

Today's kitplane era changed all that. No longer is welding considered a vital homebuilding skill; nearly all kits come with prewelded parts. No longer do builders have to laboriously assemble wooden ribs or shape metal ribs with a bar of solder; most kits supply ready-to-install ribs.

Of course, that doesn't make building easy. There's still a lot of work to do and new skills to learn. This section discusses some of the aspects of construction.

Type of construction

One factor to include when considering homebuilts is selecting which type of construction you'd like to work with. There are four basic types:

- Composite
- Metal monocoque
- Tube-and-fabric
- Wood

Composite construction works by combining two (or more) materials whose advantages complement each other and whose deficiencies cancel out. In the simplest form, "strong-but-flexible" fiberglass cloth is soaked in "stiff-but-brittle epoxy" (itself a mixture of resin and hardener) to form "strong-and-stiff" components such as cowlings and wheelpants.

The two types of aircraft composite construction are *molded* and *moldless*. The first uses a mold to define the shape of the structure. It's most suited to a production line because during construction the various components can be bonded together just like a plastic model. With moldless composite construction, the builder shapes the design from styrofoam and fiberglasses; it's more time-consuming, but cheaper.

Metal monocoque construction is used by aircraft manufacturers from Piper to Boeing. A thin sheet of flat metal is pretty flimsy, but when the metal is rolled into a wide tube, it becomes stiff. Attach some bulkheads inside to prevent the tube from collapsing, and you've got a light, strong structure ideal for aircraft (Fig. 5-4).

Fig. 5-4. Much of the aluminum used in the strong-but-lightweight Murphy Maverick is less than $25/1000$ of an inch thick, yet it withstands more than 6 Gs.

Tube-and-fabric construction is another of the "traditional" ways to make light aircraft. The main structural shape of the fuselage is defined by a metal truss. The wing can be built in one of several methods. The spars, for example, can be solid wood, built-up wooden boxes, or metal extrusion. The ribs can be stamped metal, cut plywood, built-up shapes, or even foam. Fabric is applied to the structure (hence, the term *ragwings*), then the fabric is sealed with *dope* to produce an enclosed, streamlined shape. Commercial tube-and-fabric lightplanes include classic Pipers and the Bellanca Citabria.

Wooden aircraft are built just like balsa models. The fuselage structure consists of longerons and bulkheads glued into the proper shape. Like tube-and-fabric aircraft, the wings can be built in several ways and are either sheathed in plywood or covered with fabric.

The preceding is only a general guide; a number of airplanes use a combination of methods. Generally, though, designers keep to one mode of construction throughout. Not only does it make the parts list simpler, but it reduces the construction time because the builder won't have to learn two unrelated skills.

Types of kits

The cheapest way to build an airplane is to buy a set of plans and build it entirely from scratch. That route isn't all that common anymore. The kit homebuilt, or *kitplane*, has taken over much of the homebuilt arena. There are some good reasons for this progression.

First, while building a kitplane will invariably cost more than starting from scratch, the basic price is at least known in advance, somewhat reducing financial uncertainty. Second, kit companies usually have good builder support. Large manufacturers like Stoddard-Hamilton provide a telephone helpline to assist Glasair builders. Third, a successful kit becomes a marketing tool to build further interest. If a prospective builder is wondering if an ordinary person can build a plane, a line of Van Grunsven RV-4s and RV-6s at a local fly-in is a pretty powerful answer.

All kits are not created equal, though. The amount you pay depends upon how much work is done for you. The *materials kit* contains the raw materials necessary for a plans-built aircraft. No work is performed by the kit supplier; the supplier often has no connection with the aircraft designer. Basically, it's a painless way of obtaining the necessary hardware. The kit provides aluminum sheets, plywood, metal extrusions, and other materials that require considerable work on the part of the buyer to become aircraft parts (Fig. 5-5).

The next step upward is designs that can be built from plans, but offer *subkits* to hasten construction. This is the most flexible method because the builder can buy whatever subkits her budget allows and build the rest of the airplane from scratch (Fig. 5-6). A builder unsure of his skills can buy subkits for crucial or complex components.

When the entire aircraft can be ordered as a single item, it becomes a *complete kit* (Fig. 5-7). Few kits are truly "complete" because even the best don't include paint or batteries.

Construction cost

The materials kit is the cheapest because nothing is done for you. The most expensive kits are usually those for molded composite aircraft. Fiberglass and epoxy are that costly; in addition to the materials, the buyer pays for the factory labor that was necessary to fabricate the parts.

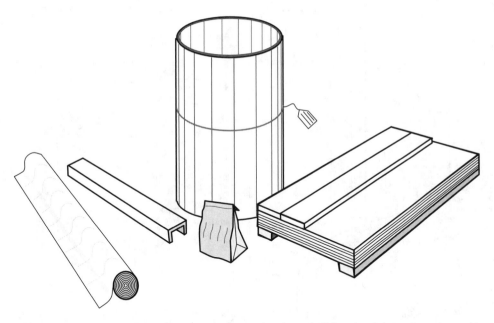

Fig. 5-5. A materials kit is the cheapest way to get into homebuilding, but it leaves you the most work!

However, the cost of the kit isn't the only factor involved. No kit includes *everything* required to build the plane. There's some additional hardware to buy, from penny-ante stuff to big-ticket items.

The biggest of the big-ticket items, of course, is the engine. Only designs using low-cost two-stroke engines include the powerplant. The additional cost is frightening for planes using conventional aircraft engines.

Engine prices can range from a few hundred dollars for a used two-stroke to well over $20,000 for the larger engines. The kit for the Van's Aircraft RV-6 sells for about $10,000, and includes just about everything needed except the engine, prop, instruments, and tires. A new Lycoming O-320 engine will cost about $14,000.

The rest of the additional items aren't quite as bad. Propeller? Pay $250 for a Rotax prop, or thousands for a constant-speed model for a Continental or Lycoming. Instruments? At least $500 for basic VFR, if you're buying new. Avionics? No kit includes radios or a transponder. Simple VFR birds can get by with just a handheld transceiver, but even these sell for more than $300. By the time you add a navigation receiver, loran, and a transponder, avionics' cost is well past $2,500.

That's not all, either. How well equipped is your workshop? Whatever tools you own, they probably aren't enough. Drill presses, band saws, table saws, grinders: It all adds up. Some kits claim that they "can be built with simple hand tools," but you'd be better off getting at least a drill press and an aluminum-cutting band saw.

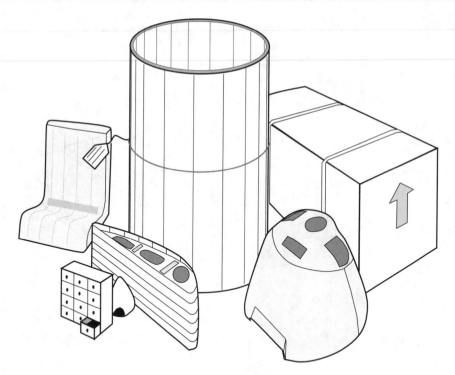

Fig. 5-6. Building aircraft under the subkit method gives the builder the most flexibility. He or she can save money by doing most of the work or buy prefabricated components.

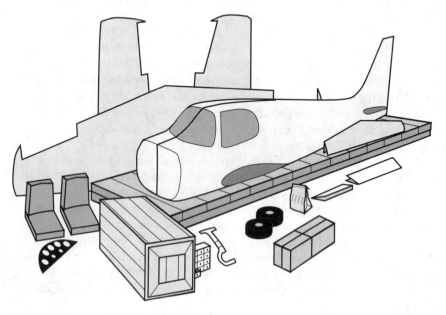

Fig. 5-7. The complete kit is the fastest to build, but the most expensive.

Construction time

All kit companies include a typical construction time with their advertising. Most builders laugh at these estimates. Generally speaking, you should double or triple the manufacturer's prediction.

If you want to spend the least time building, pick less-complex aircraft. The RV-6, for instance, has a simple, rugged, fixed gear. The Lancair has a mechanically operated retractable tricycle system. The time that a Lancair builder gains through composite construction might well be lost when it comes time for gear installation.

Of course, the kitplane manufacturer can reduce the effect of additional complexity by including extensive prefabricated components. If an assembled gear linkage can be supplied, the time impact of retractable gear is reduced.

There is a constant factor in all kits, one that determines the rock-bottom time the kit might take to build. Whether you're building a Lancair or a Kitfox, an RV or a RANS Chaos, you'll spend most of your time working on the internal systems. All airplanes have control systems to install and trouble-shoot; all have wheels and brakes to install; all must be painted.

It's fun to build primary structure, making something that *looks* like an airplane, but most of the time you're standing on your head in the cockpit or crouching in front of the firewall working on some subsystem or another. Figure 5-8 shows some of the systems that must be built and tested on most homebuilts.

Construction summary

Most kits proudly state, "Can be built with ordinary tools by someone with average skills." What's an ordinary tool? What's an average skill level? The definitions change between kit manufacturers. You can be sure by the time you're done, you'll have much higher skills and a much better-equipped shop.

It all comes at a price, of course: financially and personally. It's hard to answer all your questions in these short pages. I've written another TAB/McGraw-Hill, Inc., book, *Kitplane Construction*, for those contemplating building a kit aircraft. The book covers the selection process and basic workmanship techniques for all four types of construction.

BUYING USED HOMEBUILTS

If you can't face the construction process, but the appearance or performance of a particular homebuilt is still attractive, consider buying a used homebuilt. Prices can be quite reasonable. Some can be bought for little more than the total cost of construction.

There are a number of reasons why a homebuilt airplane comes up for sale. First, some folks like the building process more than flying. Some have built five or more aircraft and quite often are "airplane artisans" that produce excellent results. Second, the occasional builder doesn't do his homework before starting construction, and the resultant airplane doesn't meet his needs. Finally, a builder could own her creation long enough to get tired of it and wish to move on to something else.

Cockpit
- Install/fit
  - ~Doors
  - ~Seats
  - ~Carpet & upholstery
- Heat/ventilation
- Electrical system
  - ~Gauges
  - ~Lights
  - ~Circuit breakers
  - ~Switches
  - ~Avionics

- Cut and drill windshield
- Install steps and hand holds
- Seat belts and shoulder harnesses
- Control stick and rudder pedals
- Fire retardancy
- Rain proofing

Controls
- Proper rigging
- Chafe/bind elimination
- Pulleys
- Fairleads
- Cables

- Pushrods
- Access
- Flutter
- Static and dynamic balancing

Engine
- Fabricate
  - ~Exhaust system
  - ~Carburetor heat
  - ~Fuel system
  - ~Baffles
  - ~Mount for unusual engine
- Install
  - ~Engine
  - ~Electrical system
  - ~Magnetos
  - ~Fuel lines/gascolator
- Cowling access
  - ~Install dzus, piano hinges, etc.
  - ~Oil access
  - ~Maintainability

Fuel System
- Run fuel lines
- Install valves
- "Slosh" tanks
- Venting
- Ensure adequate flow
- Calibrate gauges

Landing Gear
- Nosewheel/tailwheel steering
- Brakes
  - ~Run hydraulic lines
  - ~Connect and adjust pedals
- Shimmy elimination
- Retract linkages
- Oleos/shock absorbers
- Tires
- Bearings

Fig. 5-8. No matter how complete the kit, all kits will require extensive time-consuming subsystems work.

Aviation writer Budd Davisson has a saying: "You're better off buying a used snake than a used homebuilt . . .." Obviously, significant problems could exist with the aircraft. The structure is already closed up, making detailed inspection difficult. It's quite possible for a number of hours to be flown without a hidden flaw making itself known. In one horrible example, a homebuilt flew 14 hours without wing bolts. The mistake was discovered when the wing separated in flight.

Inspections

A prepurchase inspection is even more vital for a used homebuilt than a used production aircraft. Find someone with experience on the type of aircraft, preferably someone who has built the same model. That builder should pass judgment on the workmanship and adherence to the construction plans. While some deviations are minor, changes to the control system, rigging, or basic structure should set the alarm bells ringing.

An inspection by an A&P would also be a good idea because she is experienced in rapid assessment of airframe and engine condition. Certain components might require a specialist's inspection; Rotax-powered aircraft like the Avid Flyer would require checking at the local ultralight center for someone who knows 2-stroke engines.

The aircraft should be test flown by yourself and the experienced builder you brought, rotating turns in the left seat with the owner/builder in the right seat for both flights. You can judge if the plane is right for you; the experienced builder will determine if the plane flies like it should. Carefully examine the logbooks. An aircraft that is flown regularly probably flies well (Fig. 5-9); a plane that sits a lot might have problems.

Fig. 5-9. If the owner flies the airplane a lot, it probably doesn't have any handling or safety problems.

Single-seat airplanes present a problem. Will the owner let you test fly it? There are a number of cases of prospective buyers crashing single-seaters. If your experienced builder-friend arrives in her own version of the same model aircraft, the owner shouldn't have too many worries. Or if the aircraft is otherwise acceptable, you might be able to buy the aircraft contingent on an acceptable test flight.

Maintenance

Maintenance is another issue. *You cannot receive a repairman certificate; you didn't build the aircraft.* The original builder can perform the maintenance, if he still retains his certificate and is willing. Neither is guaranteed. If buying the plane from the original builder, a fresh annual inspection should be part of the deal. Have an outside mechanic inspect the aircraft as well, but licensed mechanics charge more for annuals than prepurchase inspections.

Don't fault the builder if she is unwilling to continue maintaining the aircraft after you buy it. The builder probably isn't a professional mechanic; the legal pressures are strong. If you buy from a friend, though, they might be willing to continue.

Any A&P mechanic is allowed to maintain and inspect homebuilt aircraft. One difference between homebuilts and production aircraft is that the A&P doesn't have to be an IA to perform an annual on a homebuilt. That'll save a bit, but if the plane has a Rotax or converted auto engine (Fig. 5-10), it might be hard to find an A&P willing to be legally responsible for its maintenance and continued operation. Check around before you buy the aircraft.

Fig. 5-10. If you're going to buy a homebuilt with a converted auto engine, there is less risk if it has a tested, commercially-sold conversion like this Northwest Aero Products Ford V-6.

One interesting point is a gentleman's agreement between the FAA and the EAA. The owner of a previously built homebuilt aircraft is allowed to maintain the airplane as long as an A&P signs off the work within the next 12 months. If you can find a mechanic willing to work on this basis, you've got the best of both worlds, authorizations that are equivalent to a repairman certificate, with an experienced eye to keep the aircraft safe.

The final analysis

As you're discussing the deal with the seller, one of the primary things you should determine, perhaps the first question, is *why the aircraft is for sale*. Does the plane just have a few hours on it? That should trigger a few red lights; maybe it handles so weird that the owner is scared of it; maybe some major mistakes made during construction have been discovered.

Examine the aircraft logs. Has the plane flown regularly? If the plane were 10 years old, I'd be a lot happier to see 1,000 hours in its logbook rather than 100. Good-handling, reliable aircraft are *flown*. A plane that has been airborne for only a few hours per year might have problems.

Probably your safest route is to know the seller in advance. I've seen a number of homebuilts come up for sale within my local EAA chapters. With my knowledge of the builders, I would have had no hesitation about buying most of the aircraft.

With the boom in kitplanes, we are going to see more and more homebuilts available on the used market. They're a definite option for those looking for something more than a typical factory plane, but without the patience or inclination to build their own.

BUYING PARTIALLY COMPLETED HOMEBUILTS

There's a thriving market in flyable homebuilts. Uncompleted kits are another kettle of worms. Buying and selling unfinished homebuilts is nothing new. It used to be said that 30 percent of homebuilt aircraft projects were eventually flown—10 percent by the first builder, 10 percent by the second owner, and 10 percent by the third owner or subsequent owners. One Fly Baby ran through six owners until it was completed, 20 years after construction started.

A flying homebuilt has already passed its most crucial test; its flying characteristics can be used to judge how well it's built. An uncompleted project? Who knows what might be wrong with it? The sixth owner of that uncompleted Fly Baby found a critical problem with the wing bracing system: the *sixth* owner. Would the previous owner have detected the problem before it was too late?

Still, there are some advantages to buying partially completed planes. The primary one is the work that you won't have to do. Wouldn't you like to pick up a completed empennage for the cost of the tail kit alone?

In addition, buying an uncompleted plans-type homebuilt can be the equivalent of buying a kit. Like to build a Volksplane? You might be able to pick up most of an airframe for less than a thousand dollars, and get it flying for a couple hundred hours of construction time and just a little more money.

Buying a partially completed homebuilt is a gamble. It might hide a deadly flaw, or it might be the best aircraft deal since the $650 Jenny. The problem, of course, is telling the difference. How can you make sure you're not getting taken?

Do your research

To begin with, know what you're buying. Check the specifications. Read pilot reports. If the aircraft type never became popular, there's often a good reason that is usually rooted in flying qualities or being difficult to build. Avoid partially completed custom designs, and stick with those that have been commercially available.

If the advertised project is a kit, knowing the exact model is vital. A homebuilt can be dramatically improved in a 10-year period. An older kitplane's performance and reliability probably does not favorably compare to the latest model. The Avid Flyer of the 1990s can be built faster and is roomier than the 1985 model. The older kit came with a Cuyuna engine; the 1990s kit has a dual-ignition oil-injection Rotax 582.

When you understand the aircraft, find someone who's built one. Ask about common problems. What construction mistakes are typical? Which are the crucial construction (and flight) operations? Ask them if they know the builder with the project for sale; get their impression of the owner's workmanship. Of course, most folks will be reluctant to slur an acquaintance's ability. Rave reviews are a good sign; a shrug and a "pretty good" can mean anything.

The examination

Properly prepared, it's time to examine the merchandise. Don't go alone. Four eyes are better than two, and if the second set belongs to a builder experienced with the design, so much the better.

Scope out the builder's shop. How much equipment was he working with? Did he have an adequate air compressor to rivet with, or some cobbled-up system using a hobby compressor and a small air tank? Is there a sturdy workbench or a cheap folding table? How did he store the kit materials? Did he lean plywood against the walls, or build a flat-storage rack? Do you see a few discount-store hand tools or a complete machinist's set?

This isn't to say a person can't build an acceptable aircraft with minimal tools and facilities, but you're looking for clues toward poor workmanship. Few homebuilders rush out and completely equip their shop from the outset. Someone with a well-equipped shop probably owned the tools prior to beginning the aircraft; their workmanship would probably be good from the start. But if the owner has a less-than-optimal compressor setup, rivets might have suffered. A flimsy worktable can result in warped components. Poor storage of plywood, steel, or sheet aluminum can ruin good material.

Speaking of the raw materials, check them over for damage and decay. For wood, look for the discoloration of rot and the flaking and checking due to drying out. The end grain of wood stock should have a coating of paint or varnish to retard the drying process.

Aluminum should be free of corrosion; look for roughness and dry powdery areas. Sheets often get lightly scratched in normal shop handling, but there is a limit. Deep scratches concentrate stresses and can cause premature failure. Slight bends and wrinkles might not affect the strength, but make it awkward to work with.

Any untreated steel is likely to have a thin coating of rust. Conscientious builders apply a coat of paint as soon as possible. Unlike corrosion on aluminum, rust can be safely removed from steel, as long as it isn't deep. If the builder has neglected to prime a steel-tube fuselage, though, derusting it will take considerable time. (Documentation and evidence that the interior of steel tubing has been properly treated would be a huge plus.)

You won't have these worries with composite materials; however, epoxies usually have a maximum shelf life. Make sure it hasn't been exceeded. Some epoxies are shipped in slightly permeable plastic containers that reduce the shelf life, as well as stink up the shop. It's a positive sign if the builder transferred the resin to metal or glass containers. Finally, make sure any fiberglass cloth is dry, clean, and still on the roll.

Checking the aircraft

Of course, your primary item of interest is the aircraft itself. Check its straightness; crouch down and squint along edges. Run your hand along surfaces and check for smoothness. Have the builder remove access panels and other items that hinder a full inspection. You'll soon understand one of the biggest frustrations in buying a partially completed project: the more finished, the harder to inspect. A closed-up metal or composite wing might cover a number of sins, yet the buyer will want a higher price due to the level of completion.

Specific areas of interest vary with the type of construction. For wood, pay careful attention to the glue joints. Find out what type of glue the builder used. The poor filling qualities of traditional glues require tight joints for strength. Modern epoxies such as T-88 do a good job of filling loose joints, but tight ones are still a sign of good workmanship.

One crucial item to check is the amount of glue in the joints. The joints should all show signs of glue oozing out when the pieces were clamped. The excess is normally wiped away, but a small fillet should be visible all the way around.

Similarly, composite layups should also be checked for starvation. Completed parts should have a rich amber color; pale patches indicate voids and delamination. An excess of epoxy isn't dangerous, but it's heavy and a sign of sloppy workmanship. The cloth weave should run straight or curve evenly around bends. All parts should be dry and clean of grease and dirt.

Partially completed aluminum aircraft require careful examination of the riveting. Examine the area around the manufactured heads for skin distortion, dents, and other signs of improper riveting procedure. Check both heads for distortion or cracking. Gauge a few shop heads; make sure they're at least ½ the rivet diameter high, and 1½ the diameter wide. Mind the edge margin because the rivets should be spaced apart at least three times their diameter, and at least twice the diameter from any edge.

Check the straightness of the rivet lines. A little waviness is normal—unavoidable for some of us—but you can get a better feel for the builder's level of workmanship. Look for scratches and dents.

Welded structures can be tricky. Factory-welded areas are generally all right, although I've seen a few bad ones. Builder welds are another matter. While it's easy enough to check the surface, it's quite possible that only the surface material was melted and jointed with the underlying metal still separated. "Penetration" of the weld is important, but difficult to prove without actually cutting through the joint. Hopefully, the builder made periodic test welds to saw apart and verify penetration.

Other goodies

In addition to the aircraft itself, don't forget to examine any other equipment included in the sale. Checking an engine can be difficult, but at least make sure it includes the required accessories, such as magnetos, alternator, starter, and carburetor (Fig. 5-11). Additional firewall-forward goodies sweeten the deal, especially an engine mount.

Fig. 5-11. If you buy a partially completed homebuilt that comes with an engine, make sure the powerplant includes all components necessary for operation.

If the engine is used, scan through the logs. The engine should have been *pickled* for storage. A special oil coats the crankcase and sump, *desiccator plugs* are installed in the cylinders, exhaust, and breather to absorb moisture, and the engine should be wrapped and stored in a dry place. If not properly pickled, the engine might be a 250-pound block of rust.

Check the plans to make sure that they're complete. Glance over the inventory of kit components. If the kit originally included parts such as wheels, tires, and instruments, find out if the owner has included them in the sale. Ask to see the owner's *builder's log*: the more detail, the better. For example, it should list exactly what brand of primer was used. That'll help you avoid compatibility problems when it comes time to paint. If she has a collection of newsletters or similar publications, ensure that they're included in the deal. Request any update documentation from the kit supplier.

Ask the owner about construction variations. Most builders make small modifications, and few of these affect complexity or airworthiness. Be leery of major mods, such as conversion to a nonstandard auto engine or a scratch-built retractable gear. There's nothing wrong with people making changes like this; however, you don't want to be stuck with someone else's poorly thought out modification.

THE DECISION

Decision time finally comes. Has the builder done a safe job, so far? Homebuilts can be remarkably tolerant of bad workmanship. As long as the structure is sound, the cosmetics don't affect the airworthiness, but it's your peace of mind as well as your life at stake. It's no fun flying a plane that you have to worry about all the time. And everyone is going to assume that those bad cosmetics are your fault.

If you decide the kit is acceptable, it's time to haggle over price. Unfortunately, there are no guidelines, and there is no "blue book" for unfinished kits. Their value is determined by too many factors, from the workmanship, to completeness, to the kit's popularity.

Don't be thrown off by the "current" kit price. The original Glasair kit sold for half of the mid-1990s' price, but was markedly harder to build and the completed airplane had poorer flight characteristics. So the starting point should be about the original cost, not what a brand-new kit might run.

How much to "pay" for the work already completed is going to be the thorniest part. If you have to redo shoddy work, the project is of less value, though it might not be politic to tell that to the owner. It boils down to how much the work is worth to you. Treat the seller's description of "X-percent completed" with a ton of salt.

For a starting point, take the original kit price (or cost of the materials), then alter it up or down as necessary depending upon workmanship and additional equipment. The actual market value is probably between 80 to 120 percent of this amount.

What to offer? It depends on the situation. Some sellers are just tired of the bother and are willing to take a much lower price; others are in no hurry and are willing to wait for what they feel is a fair price. I know someone who sold a project, then bought it back later for half of the original selling price. Each case is different. It never hurts to make a low offer and see how owner reacts.

But unless you pay peanuts, it's probably less of a deal than you think. Talking to local buyers of partially completed homebuilts reveals one common theme: They had to do more work than they thought would be necessary. As everyone will tell you, the last 10 percent of the aircraft takes 90 percent of the time.

Finally, remember that you must perform at least 51 percent of the work on the aircraft to receive a repairman certificate. A partially completed project might not allow you to qualify to do your own maintenance. Obtain a ruling from an FAA inspector prior to buying an unfinished project.

CASE STUDY:
A CLASSIC KITPLANE

Many of the nonflying public might have trouble picking out the Van's Aircraft Company RV-6 as a homebuilt aircraft. After all, RV-series planes don't look that much different from other airplanes. They're not tiny; they're not canard pushers; and they're generally not painted in wild sunburst paint schemes. Most of all, they're not rare.

The RV line is one of the most popular kitplane series in the world, and the number of *flying examples* exceeds some factory aircraft's production runs. There are 35 of them flying in my local area; 14 are based at one uncontrolled airfield alone. I've seen more than 30 of them at one local fly-in.

John Ammeter (Fig. 5-12) didn't build his RV-6 because it was popular, though. He built it because no one made the airplane that he wanted to buy. He'd

Fig. 5-12. John Ammeter, RV-6 builder.

become frustrated with renting. "I'd call them up on the weekend and ask for a Cherokee, and they'd say, 'We've got a 150 between 3:30 and 5:00.'" He was ready to buy, but a production plane wouldn't do. "What I liked, I couldn't afford. What I could afford, I didn't like."

He sat down to determine his mission and the features that his dream airplane should have. He wanted speed and the ability to fly from short grass fields. He wanted a low wing, side-by-side seating, and a real aircraft engine, not a two-stroke or Volkswagen. And with his background in Cessnas and Cherokees, his dream plane would have tricycle gear. "By the time I got done," he said, "I had designed the RV-6A, the trigear version of the RV-6."

Only one problem: At the time, the RV-6A was only a gleam in Dick Van-Grunsven's eye. John decided to join the ranks of the taildragger pilots and ordered his first RV-6 subkit.

Building

The building experience went rather smoothly. He started in February 1987. There were a lot of RV aircraft being built locally, so he profited from the advice and supervision of builders farther along in the process. He bought a used Lycoming O-320 engine for $3,500. Getting it into shape cost another $500 or so.

It was important for John to stay on good terms with his wife, Sue. Therefore, he worked on the plane only two or three evenings a week, and one day every weekend. The other weekend day and the rest of the evenings were left open for family activities. Still, he worked on it steadily.

He flew N16JA (Fig. 5-13) for the first time after about three and a half years of construction. He'd planned to spend about $15,000; by the time he'd had the plane painted professionally, Ammeter had a total of $25,000 into the aircraft. While he didn't closely track the time, he estimates that he put in between 2,000 and 2,400 shop hours.

Ownership experience

"I thought I was going to make a lot of long trips. Then I realized I didn't know anybody in Oklahoma, or wherever." With the RV, and with his introduction to the world of homebuilding, that has changed. "I've enjoyed building and flying, but especially the people I meet. The RVers and EAAers are the finest folks you'd ever know." He now knows someone from Oklahoma, a fellow RV builder, of course. Charlie Calivas flew his RV-6 to Seattle to visit John and his wife. He made the trip in 8 hours. The RV certainly is fast! John has taken a number of long trips to visit family and new friends.

John met more RV enthusiasts when he attempted the usual homebuilder's triumphant trek to the Oshkosh fly in. He had to make the last few miles without N16JA. While taxiing across the Eau Claire, Wisconsin, airfield on the way to Oshkosh, the plane in front of John stopped unexpectedly. Ammeter slammed on the brakes, and the RV tipped forward. The prop shattered on the pavement.

Fig. 5-13. Ammeter's RV-6 cruises at 190 mph with the same engine as a Cessna 172.

John bit back the disappointment. He left the plane in Eau Claire and rented a car for the drive to Oshkosh. After a few days of meeting other RV builders, he returned with a replacement propeller.

There he found another example of the fine people involved in the home-building movement. A man named Perry Kuzner had noticed the kitplane sitting with a splintered prop. He'd checked the airplane every day. And when John showed up with the prop, Kuzner introduced himself and took time off from work to help get the RV airworthy again.

The broken prop has been one of the few maintenance items that the builder has had to contend with since finishing the airplane. Upon completion of the aircraft, he received the repairman certificate and now performs all maintenance and inspections himself. Parts and equipment run him a paltry $200 a year or so.

The Lycoming is approved for autofuel, and he buys it when possible. The plane is flown about 60 hours per year.

Ammeter carries only liability insurance. "If I crash it bad enough to total it, I'll probably be dead," he says. "If it isn't totaled, I'll salvage it myself and build another one." He pays $290 a year for liability coverage.

Rent on a fully enclosed hangar costs $165 per month, and the hangar is 40 miles from home; however, he's having a hangar built at a closer airport. It's a condominium arrangement. He's taken out a mortgage for the $24,000 that the hangar costs; plus he will be paying about $55 a month in dues. Figure 5-14 shows a summary of John's operating expenses.

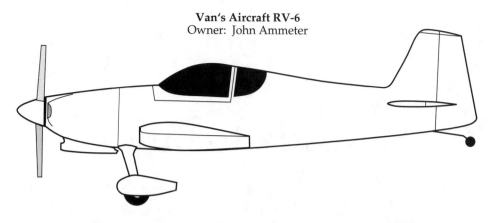

Van's Aircraft RV-6
Owner: John Ammeter

Hangar: $165/month
Insurance (liability): $292/year
Hours/year: 60

Typical annual inspection: $0
Other yearly maintenance: $200

Aircraft constructed by owner

Fig. 5-14. A summary of Ammeter's ownership expenses.

Advice

When John Ammeter started construction, he didn't know anything about building airplanes. He has become an EAA technical counselor and the head of the local RV-builder's group, the Puget Sound RVators. When asked about home-building, Ammeter replies, "It's the most fun, fulfilling, and challenging thing you'll ever do. You'll get frustrated and mad, but when you get done, you're so proud of yourself you'll hardly know what to think."

As for the RV itself? "It's the best value for the money." If you're interested in RVs or just want to talk about homebuilts in general, John would be happy to hear from you:

John Ammeter
3233 NE 95th St.
Seattle, WA 98115

6

Ready to buy

YOU WERE READY TO BUY three chapters ago, weren't you? You know your mission, and you have picked out what kind of airplane or airplanes would meet your requirements. It's time to buy.

DETERMINING THE DOLLARS

You have a pretty good idea what type of airplane to buy. The next step is to determine which models fall within your price range.

Finding the prices

With used cars, determining typical selling costs is relatively easy. One can scan the classified ads and keep a running mental tally of the price range. Or go to the bank and thumb through their blue book of automobile prices. You can, to a significant extent, do the same things with airplanes; however, there are a few differences that get in the way.

First, your local newspaper is not likely to have a wide variety of aircraft for sale. If you wanted to determine what 1985 Pontiacs sell for, you're in luck. But there are just not enough aircraft-for-sale listings in your typical newspaper to determine a price range for a given model.

Fortunately, there are a number of aviation publications with extensive "For Sale" sections (Fig. 6-1). The granddaddy of them all is *Trade-A-Plane*. This tabloid is published three times a month and includes aircraft ads for all around North America. Many consider it the standard for determining aircraft prices. It's available at most aviation supply stores. For a subscription, contact:

Trade-A-Plane
P.O. Box 509
410 West 4th St.
Crossville, TN 38557

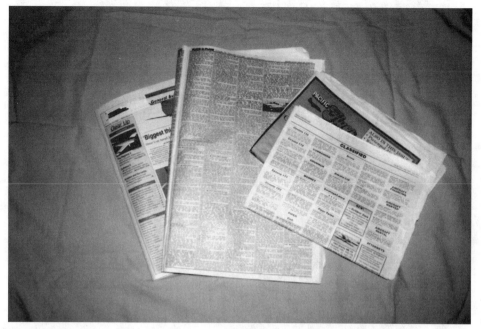
Fig. 6-1. You can get a feel for prices through any number of publications.

Several companies publish aircraft price guides. Most of these aren't sold to the general public, but you can often get access though a local sales office.

One publication is available to the general public. N.A.D.A., the company that publishes the familiar orange-cover automobile, boat, and motorcycle price guides, also sells the same sort of book for aircraft. The *Retail Aircraft Appraisal Guide* (Fig. 6-2) covers a wide variety of aircraft, including some built before World War II. Instead of using the wholesale/retail values format of the auto book, the N.A.D.A. aircraft guide includes low, average, and high prices for each year.

They gather the information from a wide variety of sources: sales reports from private owners and dealers, insurance companies, finance companies, state tax offices, appraisers, advertisements in metropolitan newspapers, and publications such as *Trade-A-Plane*. The format includes adjustment tables for factoring in engine hours and type, as well as a number of useful cross-reference tables.

It's a slick little book, which they sell to the general public. To order, call 800-966-6232. A one-year subscription, three issues, is $85; single issues are $42.50 (prices subject to change).

The want-ad trap

There is a rather insidious hazard involved with reading aircraft-sales ads. Say you're interested in a Cessna 172 in about the $20,000 range. You'll read through the ads and find a couple in your price range.

Fig. 6-2. Sales of the *N.A.D.A. Retail Aircraft Appraisal Guide* are not limited to dealers.

But then you'll stray to another category. Hmmm, here's a '63 Mooney for only $25,000, or a Cessna 310 twin for only a couple of thousand more. Not so very long ago, I found a DC-3 in supposedly flyable condition for $38,000!

Chapter 3 devoted a lot of time and effort to establishing a *mission* and the least aircraft that can meet mission requirements. That undertaking will help you avoid falling into the traps that these ads can set.

The operating cost of a used airplane is not based on its purchase price. The operating cost is determined by how much the plane would sell for *if it were new*. If you had $40,000–$50,000, you could buy a 1980 Piper Warrior or a 1960 Beech Bonanza. But a new Bonanza might cost three times more than a small four-seat Piper.

The parts price will reflect the same ratio. It's actually not that different from cars; an acquaintance paid $300 for an ordinary tune-up on a 20-year-old Mercedes, and I rebuilt my VW's engine for the same money. Avoiding this trap is simple: Pick a plane and stick with it.

The exception

You probably won't go far wrong with this advice; however, in the real world, it's not a bad idea to have an alternate aircraft in mind. If you're looking for a Cessna 182, for instance, remember that a Piper Archer's performance and fea-

Fig. 6-3. If you're in the market for a Cessna 182, keep on the lookout for good deals on alternative choices, like Cessna 180s or 185s.

tures are somewhat similar. Or maybe a taildragger equivalent might fit your needs (Fig. 6-3). Do a little pricing on an alternative or two, just to be able to tell if a good deal arises unexpectedly.

Why? Maybe, just maybe, you'll latch onto a bargain. Unless you've got the hots for a particular make and model, the ability to change targets might prove useful. By doing your homework on a couple of alternatives in advance, you might be able get more bang for the buck.

Financing

The primary operative at this point is how much money you can afford to spend. If ready cash is burning a hole in your pocket, this part is relatively easy. However, if you plan on financing, it's best to lay the groundwork prior to starting your search. A seller might not want to wait while your application works its way through a financial institution.

I had a prime example of this when I sold my 150. One younger man came by for a test flight. He was a first-time buyer, and he was very enthusiastic about my little red-and-white bird, but he didn't have the money available. He asked if I would hold the plane for him. I declined because I figured that it would sell relatively fast. He said that he'd give me a call when he got the money together.

Sure enough, he called six weeks later, a week after the plane was sold. He was crushed, and I'll be honest: It hurt me as much as it did him. I don't know why it took him that long to raise the money. Maybe he had to sell a car, first. But if he'd handled the money details *before* he started looking, he'd have made us both happy.

As far as financing, you have three choices: a loan with the aircraft as collateral; a loan with another item of property for security; or an unsecured loan, sometimes called a "signature" loan.

You can finance an airplane just like a car, with the vehicle itself acting as collateral. Maximum loan periods will range from 5–15 years, depending upon the age of the aircraft and the amount financed. Seven years seems to be the typical maximum for most small-aircraft loans. You can go shorter, of course.

Expect to put about 20 percent down on the aircraft. Finance rates seem to run about the same as car loans. I was quoted rates from 1.5 percent to 4 percent over the prime rate on a fixed-payment loan. Contact the company in advance to lay the groundwork; most companies can close a loan within 48 hours of your finding your dream plane.

Potential borrowers might encounter several problems. Many companies limit the age of the aircraft that they will finance. In most cases, 25 years is the maximum. With the average lightplane being about 22 years old, it's obvious that finding loans is going to be tough for almost half of the general aviation fleet.

The other problem is that some institutions won't make aircraft loans for less than $25,000. This leaves out a number of aircraft, especially most two-seaters. The reason for this limitation is interesting. Loans of less than $25,000 are considered to be *consumer loans,* and consumer-protection laws kick in, topped off with much more paperwork. As one financial officer told me candidly, "We have to treat the customer as an unsophisticated borrower." Ouch!

At least there are one or two rays of hope. AOPA has a program set up with Maryland National Bank to meet aircraft-loan needs. The bank's minimum loan amount is $10,000, and it will handle older planes.

Local banks might be willing to handle your purchase. Recall Claude Abbott's Bonanza in chapter 3; he took out a $35,000 loan with a neighborhood lender and later added to it to finance his Bonanza's engine rebuild.

My discussions with several local banks and savings and loans indicated that aircraft loans weren't unknown to them. One bank manager was quite positive about lending money for airplanes. Another local bank advertises that it will loan to a minimum of $3,500 on terms of 72–120 months, based on the loan amount.

Of course, one needn't take out a loan on the aircraft itself. A lot of planes are financed through loans secured in other ways, such as a home equity loan. In addition, you can get unsecured loans based solely on having good credit. My credit union has preapproved signature loans for up to $20,000. The interest rate is a bit higher, though.

These types of unsecured loans have one big advantage over the traditional aircraft loan: Aircraft loans require a 15–25 percent down payment. You can borrow the full amount necessary, and keep your cash for a maintenance reserve. Or even borrow $2,000 or so over the cost of the plane, to cover immediate expenses.

A second advantage of these unsecured loans is that you won't be required to carry full insurance on the aircraft. Such insurance is probably a good idea, if you're deeply in debt, you don't *have* to add it to your aircraft budget.

GET SET

Plane's picked out, money's settled: You have just a couple more things to take care of before leaping into the market.

Knowing the warts

No designer is perfect. All airplanes have revealed certain problems over time, and problems equal money, as far as the aircraft owner is concerned. As mentioned in chapter 3, *Aviation Consumer* performs in-depth reviews of various small aircraft. These reviews include a good summary of the aircraft's maintenance tendencies. You'd like to know what these tendencies are, so when you're looking at a prospective purchase, you can concentrate on these trouble areas. The magazine's reviews have been collected and published in *The Aviation Consumer Guide to Used Aircraft*.

Serious aircraft mechanical faults are the subject of FAA airworthiness directives (AD). Compliance with ADs is mandatory, yet a few owners seem to sneak by. You sure don't want to buy their plane and end up paying for the missed ADs. A little research is in order.

You can research the applicable ADs at an FAA flight standards district office or the base of an agreeable A&P with inspection authorization (IA). IAs are required to maintain an AD file, but they aren't obligated to let the general public paw through their listings. I talked to a couple of FBOs who said they wouldn't be too adverse to giving access to their AD list. If your local IAs are agreeable, don't abuse their good nature. And give them some of your maintenance business after you buy.

The place to start is with the index. This publication lists the ADs issued against aircraft, engines, propellers, and what the FAA calls *appliances* (magnetos, hoses, etc.), grouped by manufacturer and model. A sample page from this index is shown in Fig. 6-4. The index includes the AD number and a short description of the problem.

When using the index, first look under the manufacturer's name, then find the model of aircraft that you're interested in. There'll be two listings of interest. The first is the *series* heading. This leads a set of ADs that apply to the entire series of aircraft, no matter the year or model number. After the series heading, the individual models and years of the aircraft type are listed. Both sets of ADs apply to your prospective purchase. The listings under the aircraft type include only those ADs that apply specifically to the airframe. There are separate sections for the engines, propellers, and appliances.

You might not know what engine and propeller to look at. Sure, you'll know a mid-1970s 172 has a Lycoming O-320, but the O-320-*D2J* and O-320-*H2AD* were used. You can find out exactly which engine and propeller a particular aircraft used by consulting its *type certificate data sheet*. A sample page of a data sheet is shown in Fig. 6-5. The FSDO and your friendly IA will have copies of these, as well.

Getting back to ADs, each issuance is identified by a group of three numbers, such as 82-10-03 or 64-20-01. The first two digits are the year in which the AD was published. The second indicates in which "series" the AD was published; these series are generally released biweekly. Finally, the last two digits indicate where this AD was included in the series.

With a list of the ADs applicable to your prospective purchase, what's your next move? You *could* look up the ADs to find out what they cover. There might be

Cessna (continued)

MAKE & MODEL	TC NO.	AD NUMBER*		SUBJECT	PAGE NO.**	BOOK NO.
140	A-768	46-44-01		Rudder stop bolts	1	1
		46-44-02		Safety belt bracket reinforcement	1	1
		46-44-03		Windshield retaining channel	1	1
		46-44-04		Carburetor hot air ducts	2	1
		46-44-05		Engine mounting bolts	2	1
		47-06-10		Aileron carry-through bar	3	1
		47-06-11		Forward doorpost cracks	3	1
		47-26-02		Wing leading edge rework	3	1
		47-43-01		Primer line relocation	4	1
		47-43-02		Fuel selector valve handle	5	1
		47-43-03		Seaplane spreader struts	5	1
		47-43-04		Rudder control cable horns	5	1
		47-43-05		Elevator spar web reinforcement	6	1
		47-43-06		Aileron support ribs	6	1
		47-43-08		Beech R003-201 propeller blades	6	1
		47-50-02		Fuselage bulkhead	6	1
		48-05-04		Operator limitations placard	6	1
		48-07-01		Stabilizer attaching bolts	7	1
		48-25-02		Welded exhaust muffler	7	1
		48-25-03		Wing drag wire system	7	1
		50-31-01		Fin spar reinforcement	8	1
		51-21-01		Rudder rib flanges	9	1
		61-25-01		Met-Co-Aire landing gear	12	1
		62-24-03		Cabin heat system	13	1
		79-08-03		Electrical system	83	1
		86-26-04		Shoulder harness adjuster	51	2
140A	5A2	61-25-01		Met-Co-Aire landing gear	12	1
		62-24-03		Cabin heat system	13	1
		79-10-14	R1	Fuel tank venting	84	1
		86-26-04		Shoulder harness adjuster	51	2
150 Series	3A19	62-22-01		Vacuum pump modification	12	1
		67-03-01		Exhaust gas heat exchangers	14	1
		68-17-04		Stall warning system	15	1
		71-22-02		Cracks in nose gear fork	24	1
		79-08-03		Electrical system	83	1
		79-10-14	R1	Fuel tank venting	84	1
150	3A19	73-23-07		Defective spar attach fittings	35	1
		74-06-02		AVCON mufflers	36	1
		75-15-08		Engine lubrication	40	1
		77-02-09		Wing flap system	63	1
		86-15-07		Modification-larger engine	44	2
		86-26-04		Shoulder harness adjuster	51	2
		87-21-05		Placard-spins	59	2
		88-15-06		Battery location	66	2
150A	3A19	75-15-08		Engine lubrication	40	1
		86-15-07		Modification-larger engine	44	2
		86-26-04		Shoulder harness adjuster	51	2
		87-20-03	R2	Seat tracks	53	2
		87-21-05		Placard-spins	59	2
		88-15-06		Battery location	66	2
150B	3A19	75-15-08		Engine lubrication	40	1
		86-15-07		Modification-larger engine	44	2
		86-26-04		Shoulder harness adjuster	51	2
		87-20-03	R2	Seat tracks	53	2
		87-21-05		Placard-spins	59	2
		88-15-06		Battery location	66	2

```
*  T - telegraphic, L - letter, R(1) - indicating first revision since issue.
** The page numbers listed are for the sections designated by the manufacturer's name.
```

Fig. 6-4. This is a sample page from the *Small Aircraft and Rotorcraft Airworthiness Directive Index.*

a subject or two there that piques your interest; if the subject were "Wing Attach Bolt Failures," you'd undoubtedly want to find out more—and jolly quick!

One item of interest is whether the AD is *one-time* or *recurring*. An owner's involvement with one-time ADs ends when the steps described in the notice are performed and logged; a typical example is the replacement of a flawed com-

DEPARTMENT OF TRANSPORTATION
FEDERAL AVIATION ADMINISTRATION

A13CE
Revision 22
CESSNA
177
177A
177B
September 15, 1985

TYPE CERTIFICATE DATA SHEET NO. A13CE

This data sheet which is part of Type Certificate No. A13CE prescribes conditions and limitations under which the product for which the type certificate was issued meets the airworthiness requirements of the Federal Aviation Regulations.

Type Certificate Holder	Cessna Aircraft Company P. O. Box 7704 Wichita, Kansas 67277

I - Model 177, Cardinal, 4 PCLM (Normal Category), Approved February 16, 1967
2 PCLM (Utility Category), Approved August 8, 1967

Engine	Lycoming O-320-E2D
*Fuel	80/87 minimum grade aviation gasoline
*Engine limits	For all operations, 2700 rpm. (150 hp.)
Propeller and propeller limits	McCauley 1C172/TM Diameter: not over 76 in., not under 74 in. Static rpm. at maximum permissible throttle setting: not over 2360, not under 2260 No additional tolerance permitted
*Airspeed limits (CAS)	Never exceed 185 mph. (160 knots) Maximum structural cruising 145 mph. (125 knots) Maneuvering 113 mph. (98 knots) Flaps extended 105 mph. (91 knots)
C.G. range	Normal category: (+101.0) to (+114.5) at 2000 lb. or less (+105.5) to (+114.5) at 2350 lb. Straight line variation between points given Utility Category: (+101.0) to (+109.9) at 2000 lb. or less (+103.6) to (+109.0) at 2200 lb.
Empty wt. C.G. range	None
*Maximum weight	Normal category: 2350 lb. Utility category: 2200 lb.
No. of seats	4 (2 at sta. +93.0, 2 at sta. +134.0)
Maximum baggage	120 lb. (+162.0)
Fuel capacity	49 gal. (two 24.5 gal. fuel bays in wing at sta. +112; 48 gal. usable) See Note 1 for data on system fuel
Oil capacity	8 qt. (+44) (2 qt. unusable) See Note 1 for data on undrainable oil

Control surface movements			
	Wing flaps		Down 30° ±2°
	Aileron	Up 20° ±2°	Down 15° ±2°
	Stabilator	Up 20° ±1°	Down 5° ±1°
	Stabilator tab	Up 2° ±1°	Down 7° ±1°
	Rudder(measured perpendicularly to hinge line)	Right 24° ±1°	Left 24° ±1°

Page No.	1	2	3	4	5	6	7	8	9	10
Rev. No.	22	22	21	20	20	20	20	21	20	21

Fig. 6-5. The type certificate data sheet defines the exact configuration of each model of aircraft. Each data sheet is actually several pages long.

ponent. A recurring AD must be performed at intervals based upon aircraft or calendar time: "Inspect the bellcrank using a dye penetrant every 200 hours," for example.

You're probably going to be more concerned with the recurring ones than the one-timers. You'll check that the one-time ADs have been complied with; if so, they won't cost you anything.

On the other hand, looking up the AD text at this point might just be wasted effort. After all, the mechanic performing the prepurchase inspection on your prospective aircraft will be checking the same paperwork. All you really need is a listing of what ADs should have been done so that you can weed out poorly maintained candidates. The index supplies the AD number.

Then again, if you are at the bookshelf or microfiche projector and have the time, indulge yourself. There's a lot of interesting stuff in the AD listings. If you haven't decided on a particular model yet, the AD listing might aid the decision. A sample AD is included as Fig. 6-6.

Service bulletins and *service difficulty reports* are also of interest. Service bulletins are published by the aircraft manufacturer. They're similar to airworthiness directives, but complying with them is not required by regulation, though the manufacturer might term them "mandatory." Often they're the precursor to an AD. An IA should have these bulletins, and the FAA's microfiche AD records includes service bulletins as well.

Whenever A&Ps discover unusual problems during maintenance, the FAA encourages them to file service difficulty reports (SDRs). These are compiled and published periodically. A typical SDR might call attention to an abnormal wear pattern or a peculiar anomaly in a particular aircraft. Often they include a recommended method for preventing or solving an occurrence on other aircraft. SDRs aren't mandatory either; however, the FAA tracks them to try and detect patterns. Often an AD results if a certain problem occurs in a number of aircraft.

Obviously, researching the ADs, service bulletins, and SDRs can be time-consuming. It's kind of fun, sort of like playing aviation detective, but if your fantasies favor the role of Chuck Yeager over that of Hercule Poirot, there are several easier ways of obtaining this data. All it takes is, you guessed it, money.

A number of companies will perform the AD search for you. Prices range from $25 on up, depending upon the degree of information requested; however, they are mostly tuned toward providing the search for a particular aircraft, not a range of aircraft models such as Cessna 172Bs through 172Es. Call and ask. Some do provide the more general search services. All providers are quite happy to take your credit card number over the phone and mail the results to you in a business day or less, or by fax within an hour. Prices vary; get an exact quote prior to reading off your credit card number.

Providers of these services advertise in various airplane publications. Here are three:

- AD Information Services 800-952-9082
- Aerotech Publications 800-235-6444
- AOPA Title and Escrow Services 800-654-4700

67-03-01
CESSNA: Amdt. 39-338 Part 39 Federal Register January 14, 1967. Applies to Model 150 Series Airplanes, Serial Numbers 17001 through 15061328, Equipped With Right-Hand Exhaust Gas Cabin Heat Exchanger (Muffler).

Compliance required within the next 50 hours' time in service after the effective date of this AD, unless already accomplished within the last 50 hours' time in service, and thereafter at intervals not to exceed 100 hours' time in service from the last inspection, until the exhaust system is modified in accordance with Cessna Service Letter No. 65-72 dated July 27, 1965, or later FAA-approved revision, or an equivalent approved by the Chief, Engineering and Manufacturing Branch, FAA Central Region.

Several cases of exhaust gas cabin heat exchanger cracking have occurred, allowing carbon monoxide to enter the cabin with cabin heat "ON." To minimize the possibility of carbon monoxide contamination of cabin air, accomplish either of the following or an FAA approved equivalent:

(a)   Inspect right-hand exhaust gas cabin heat exchanger (muffler) for cracks by conducting a pressure test of 1 1/2 p.s.i. in accordance with paragraph 12-93 of Cessna 100 Series Service Manual dated November 1962. Replace cracked exhaust gas cabin heat exchanger before further flight with an exchanger inspected in accordance with this AD and found free of cracks.

(b)   Conduct a ground test with a carbon monoxide indicator by heading the airplane into the wind, warming the engine on the ground, advancing throttle to full static r.p.m. with cabin heater "ON", and taking carbon monoxide readings of the heated air stream at the cabin heater deflector, P/N 0411824, on the firewall inside the cabin and another reading immediately thereafter in free air 15 feet in front of the propeller, with engine shut down. If carbon monoxide indication in the cabin is greater than in free air, comply with (a) before further flight.

This supersedes AD 64-17-03.

This directive effective January 14, 1967.

Fig. 6-6. This sample AD note specifies the applicable aircraft model and serial numbers, the compliance period, and the work to be performed.

Before you call, determine the years and model numbers of the aircraft that you're interested in. Find out the exact engine models used, ditto the propeller and appliances, like the magnetos. You can chase down most of this data from an aircraft owner's manual. If you can't find the exact model numbers for the prop (especially if it's fixed-pitch) or appliances, don't sweat it. The A&P should determine the exact applicable ADs during the prepurchase inspection.

Knowing the words

When you learned how to fly, you had to learn a new language. Say "Ground point niner off the active" to someone at the bus stop, and you'll end up with a seat all to yourself—works great. It's no different when buying aircraft. There's a whole new terminology of abbreviations and shorthand to absorb.

Table 6-1 defines a set of common aircraft-sales terms and abbreviations. Some require a bit more explanation:

Table 6-1. Aircraft advertising terms and abbreviations.

ADF:	Automatic Direction Finder
AFTT:	Airframe Total Time
AP or A/P:	Autopilot
Auto Gas STC:	Approved to run on car gas
CH:	Channel (i.e., "720 CH comm radio)
CS or C/S:	Constant Speed propeller
DME:	Distance Measuring Equipment
EGT:	Exhaust Gas Temperature instrumentation
Enc:	Transponder altitude Encoder
FD:	Flight Director
GPS:	Global Positioning System (satellite navigation) receiver
GS:	Glide Slope
HC:	Heavy Case*
HSI:	Horizontal Situation Indicator
MB:	Marker Beacon
MDH:	Major Damage History*
N/C:	Nav-Comm radio
NDH:	No Damage History*
NMDH:	No Major Damage History*
NRDH:	No Recent Damage History*
RNav:	Area Navigation equipment (Generally obsoleted by LORAN and GPS)
720:	720-channel communications radio
SCMOH:	Since Chrome Major Overhaul*
SFRM:	Since Factory Remanufactured Engine*
SMOH:	Since Major Overhaul*
SPOH:	Since Propeller Overhaul*
STC:	Supplemental Type Certificate
Stits:	Covered in Stits-brand fabric
STOH:	Since Top Overhaul*
TBO:	Time Between Overhauls
STOL:	Short Takeoff and Landing modifications
TT:	Total (flight) Time
TTAE:	Total Time Airframe and Engine
TTAF:	Total Time on Airframe only
TTSN:	Total Time Since New
XPDR, Xponder:	Transponder

*Further explanation in text

Damage history (NDH, NMDH, MDH, NRDH)

This indicates whether the aircraft has ever required airframe repair for any reason, from a flight accident to a hangar door falling on it.

No damage history (NDH) is easily understood. *No major damage history* (NMDH) indicates that some slight repairs had been necessary at some point. For a metal airplane, this usually indicates that some metal had to be repaired or replaced.

Major damage history (MDH) means the plane got seriously bunged up at some point. Listing a plane as MDH in an ad takes a certain amount of guts; there's obviously something in the logs that no amount of salesmanship can hide. FAR Part 43 defines the operations that count as major repairs. Unfortunately, some mechanics and sellers have their own definitions of what constitutes "major" damage.

No recent damage history (NRDH) indicates the plane has some old repairs. It's not quite as bad as MDH because it indicates that the repairs were successful enough to allow the plane to continue flying for a considerable time. Still, it hurts the value of the airplane.

Given your druthers, you'd prefer a plane that's never felt the mechanic's rivet gun or welding torch. Given the age of the general aviation fleet, though, you might have to settle for a well-repaired bird. That's the key, of course: *well-repaired*. Such repairs are one item that your mechanic should examine during a prepurchase inspection.

Aircraft come under more scrutiny during the annual if they've ever been subjected to a major repair. Check with your local A&P if you find a hot prospect listed as MDH.

Time since overhaul (SMOH, SFRM, STOH, SCMOH, SCTOH)

The biggest cost driver on any small aircraft is the condition of the engine. Any advertisement should make some reference to the number of hours the engine has run since it was new or since its last overhaul. The manufacturers of certified engines assign a recommended *time between overhauls* (TBO) for their products.

A TBO only legally applies to aircraft in commercial service. An individual owner may fly an aircraft beyond its engine's TBO. Even so, it's a risk. The fact that the engine has a lot of running time since a detailed internal inspection hurts the engine's value and the value of the plane.

Two engine times are tracked: *total time* (TT) and *time since major overhaul* (SMOH). An engine listed as "3500 TT, 850 SMOH" has a total of 3,500 flight hours and was overhauled 850 hours ago.

You won't see the TT in two cases. The first case is if the engine has never been overhauled; "1100 TTSN" might be a typical listing (TTSN: *total time since new*). The second case is if the original engine maker has *remanufactured* the engine. An ordinary overhaul requires that parts worn past a certain point (*service limits*) be replaced. When the maker remanufactures the engine, only parts that meet or beat new-engine tolerances are kept. The original engine log is thrown away and any time that the engine accrued up to that point is zeroed. The engine on a used aircraft might then be listed as "450 SFRM," which means 450 hours since factory remanufactured. These engines are also referred to as *zero-timed*.

Independent shops can also rebuild engines to "new limits," just like remanufacturing: however, the FAA *only allows the original engine manufacturer to zero-time engines*. An independent company's "new limit" rebuild might be of higher quality than the factory's remanufacture, but the independent must still retain the engine's accumulated time.

Cylinders and valves seem to cause the most problems on aircraft engines. Often, they require replacement or a rebuild prior to the rest of the engine. A *top overhaul* restores the cylinders, pistons, and valve train to at least service limits. Prospective buyers might then see an engine listed as "2600 TT, 1200 SMOH, 10 STOH," which means 2,600 since the engine was new, 1,200 hours since its last major overhaul, and 10 since the top overhaul.

Chrome-plated cylinders are sometimes used in the overhaul process. The chrome plating increases the hardness of the surfaces, reducing wear significantly; however, the process is expensive and sometimes difficult to break in. The "C" in listings such as "400 SCMOH" or "120 SCTOH" means that the cylinders were chromed at the last overhaul/top overhaul.

Heavy case

One of the disasters that can happen to a small aircraft engine is a crack in the crankcase. At worst, the crankcase must be scrapped. At best, the engine must be totally disassembled so the crack can be welded by a specialist.

If a certain engine seems prone to crankcase cracking, the manufacturer will often design a case with thicker walls. Such engines are preferable. Your research should indicate which models of your target aircraft used the thinner-case engine; however, owners who are faced with a case repair sometimes replace the entire engine with the improved model. When the airplanes go on sale, the owners will indicate the presence of the heavier-cased engine.

The thin-walled versions aren't the kiss of death, though. One school of thought says that if the engine makes it through a full run to TBO without cracking, it probably never will. The cracking is sometimes initiated by unrelieved manufacturing stresses, and an engine that goes 1,500 hours or more without cracking probably doesn't have the stress points.

Other preparations

Get some insurance quotes based on your target aircraft. Work out with the agent or representative how to initiate a new policy. You're going to be flying your new toy a lot right after the purchase. Make sure you'll be able to get an immediate start on your insurance coverage.

If you want a hangar and haven't put your name on the list, do so now. Why did you put it off so long? The Bonanza owner in chapter 3 spent five years on a waiting list before a hangar came open at the closest field.

In the meantime, you can probably find outside tiedowns available in most areas. There's never been an airplane made that couldn't stand a month or two in an open tiedown. Figure out in advance where you're going to keep your new purchase for the first couple of weeks at least.

As mentioned earlier, you might have to pay a local tax or fee on an aircraft purchase. You might do some scouting out in advance on how to minimize this. Just taking possession in a different county might make a difference.

Finally, contact the FAA or ask an FBO for a copy or two of Form AC 8050-2, which is the aircraft bill of sale. A lot more forms will have to be filled out afterward (surprise, surprise), but the bill of sale is the only one that must be signed by the seller. You might as well pick up a copy of Form AC 8050-1 (registration application) and FCC Form 404 (radio station license application) while you're at it.

As of this point, you're ready to start actively searching.

THE HUNT

Where are you going to find your dream plane? In a perfect world, it'll be sitting on the ramp at the airport you normally fly out of. You'll be walking back from another session in a clapped-out rental airplane, and there it'll be—a faded "For Sale" sign maintains a yellowed scotch-tape cling inside the window. The model is exactly the one you're looking for. The price is right in your range, and you recall seeing it running around the pattern the previous weekend (Fig. 6-7). Why was this situation so ideal?

Fig. 6-7. Sometimes the most enjoyable way to shop for an airplane is to wander around airports looking for "For Sale" signs.

- You can take all the time you want to examine the airplane. If it's getting dark, you can come back the next day. If the plane were at an airport 200 miles away, you'd feel pressured to decide yes/no immediately to avoid another long trip.

- Because it's your home airport, you'll probably have a good idea where to find a mechanic to perform a prepurchase inspection.

- The faded sign indicates that it's been on the market for a while. It could be overpriced for its condition, or perhaps the owner doesn't want to pay for any advertising.

- If you get to the point of a test flight, you'll be operating in familiar airspace. You'll be able to concentrate on checking out the plane rather than navigating over unfamiliar ground.

- The fact that you've seen it flying is a sign that the aircraft is (probably) airworthy.

Such an ideal situation probably won't happen. Still, there are some lessons here.

Hints for the hunter

Start locally. Checking out a prospective purchase is far easier when the plane is based nearby. Wander the local airports. Many now have security fences to keep out the riffraff, but your pilot's license should be sufficient for entry. Check the bulletin boards. There's usually a board at the airport office, at any FBO, and outside the airport restaurant.

Note that such a search is a good justification for renting an aircraft on a nice Saturday afternoon. Instead of the famous $50 hamburger, you could be searching for the $15,000 Piper. Use the phone, too. Local libraries carry the phone books for nearby cities; look up the FBOs and give them a call.

Other options

There's nothing wrong with consulting the national ads like those in *Trade-A-Plane* or *General Aviation News*, especially if you're looking for a very specific aircraft or level of equipment.

Unfortunately, a lot of the aircraft will be out of convenient range. If you live in Texas and find a plane for sale in New Hampshire, what next? Unless you have a flying friend up there, you'll have to travel to New England to examine the aircraft. One wasted trip just drives up the effective price you pay for whatever plane that you eventually buy. To help narrow down your search, Table 6-2 gives an index of telephone area codes.

Buying a plane through a broker is a possibility. It's their business to know the location of available airplanes. Plus, they should be able to help you find a loan, if necessary. Most of us don't buy too many planes in a lifetime; it can be useful to get the assistance of someone who buys an airplane every week.

However, remember that a broker's income stems from people who *buy* airplanes, not folks holding out for exactly the right machine, nor those who take a leisurely approach to the hunt. Like any field, aircraft brokering has its share of un-

Table 6-2. Area codes.

Area code	Location	Area code	Location	Area code	Location
201	New Jersey	413	Massachusetts	706	Georgia
202	District of Columbia	414	Wisconsin	707	California
203	Connecticut	415	California	708	Illinois
204	Manitoba	416	Ontario	709	Newf. & Lab.
205	Alabama	417	Missouri	712	Iowa
206	Washington	418	Quebec	713	Texas
207	Maine	419	Ohio	714	California
208	Idaho	501	Arkansas	715	Wisconsin
209	California	502	Kentucky	716	New York
210	Texas	503	Oregon	717	Pennsylvania
212	New York	504	Louisiana	718	New York
213	California	505	New Mexico	719	Colorado
214	Texas	506	New Brunswick	801	Utah
215	Pennsylvania	507	Minnesota	802	Vermont
216	Ohio	508	Massachusetts	803	South Carolina
217	Illinois	509	Washington	804	Virginia
218	Minnesota	510	California	805	California
219	Indiana	512	Texas	806	Texas
301	Maryland	513	Ohio	807	Ontario
302	Delaware	514	Quebec	808	Hawaii
303	Colorado	515	Iowa	809	Puerto Rico
304	West Virginia	516	New York	812	Indiana
305	Florida	517	Michigan	813	Florida
306	Saskatchewan	518	New York	814	Pennsylvania
307	Wyoming	519	Ontario	815	Illinois
308	Nebraska	601	Mississippi	816	Missouri
309	Illinois	602	Arizona	817	Texas
310	California	603	New Hampshire	818	California
312	Illinois	604	British Columbia	819	Quebec
313	Michigan	605	South Dakota	901	Tennessee
314	Missouri	606	Kentucky	902	PEI and Nova Scotia
315	New York	607	New York	903	Texas
316	Kansas	608	Wisconsin	904	Florida
317	Indiana	609	New Jersey	905	Ontario
318	Louisiana	612	Minnesota	906	Michigan
319	Iowa	613	Ontario	907	Alaska
401	Rhode Island	614	Ohio	908	New Jersey
402	Nebraska	615	Tennessee	909	California
403	Alberta, Yukon	616	Michigan	912	Georgia
404	Georgia	617	Massachusetts	913	Kansas
405	Oklahoma	618	Illinois	914	New York
406	Montana	619	California	915	Texas
407	Florida	701	North Dakota	916	California
408	California	702	Nevada	917	New York
409	Texas	703	Virginia	918	Oklahoma
410	Maryland	704	North Carolina	919	North Carolina
412	Pennsylvania	705	Ontario		

ethical people. Some can be less than truthful about the equipment or condition of the planes they're trying to sell. Get everything in writing, and arrange your own prepurchase inspection.

Finally, while a broker can help you find the right airplane, the price will invariably be higher. And because their commissions are proportional to the value of the airplane, don't expect much help if you're in the Tri-Pacer/170 market.

Special cases

You might come across a couple of special cases during your search: ramp queens or hangar queens and airplanes sans logbooks.

Ramp or hangar queens are seen at any airport, sagging, covered with dust, obviously not having flown for years (Fig. 6-8). Why are they so neglected? Any number of reasons: The owner might have lost a medical but still holds out hope of getting it back. Perhaps money is tight, and they're just hanging onto it until times get better. Maybe there's something wrong with the plane and the owner just can't afford to fix it at present. You'll look at the plane, and a little thought will tickle the back of your brain. "Maybe I can pick it up for a song. A little TLC, and I can be flying."

Two surefire ways to hurt an aircraft are performing split-S maneuvers from 200 feet and abandonment on a ramp or in a hangar. Neglect does major, big-time damage to airplanes; engines rust internally, roughening the cylinder walls and bearings; moisture and leftover combustion products produce acids that attack metal.

Fig. 6-8. Close-up of a ramp queen. The tire is not flat; it sank into the soft asphalt. Moss grew on the underside of the wings. While you might be able to buy the aircraft for little money, getting it back into flying shape will be expensive.

It doesn't take much idleness to cause problems. The TBO on Lycoming engines, for instance, is based upon 25 hours of operation *per month*. Consider that many, perhaps most, general aviation aircraft operate only 100 hours or fewer per year.

It isn't just the engine, either. Moisture seeps into the cabin, rotting upholstery and carpets. It oozes into instruments, rusting them from the inside. Control bearings freeze. Pulleys lock.

And our feathered and furred friends get to work. At my home field, I'd watched a Globe Swift sit out in the open for years. The owner stubbornly hung onto it, the tires went flat, pieces disappeared, and the windows crazed and opaqued.

Finally, someone talked the owner into selling and began the restoration process. I was there when he opened up one of the inspection panels in the wing, reached inside, and started pulling out handfuls of bird and mouse nests. "And not one of them little critters had been plane-trained." Avian and mammalian urine and excreta will corrode anything that it comes in contact with.

That's not to say you can't make a good deal on a hangar or ramp queen. But it'll need more than just a cursory inspection before you fly it. You might luck out, or you might end up with a nine-year restoration, like Mike Furlong in chapter 4.

The second type of special case that you'll run into is an aircraft without logbooks, or a single logbook for airframe or engine is missing. The logs might have been stolen, or perhaps the aircraft was seized from a drug smuggler or dealer and is up for a government auction.

Missing logbooks is a sticky situation. You won't know the time on the airframe or engine. You'll have utterly no idea about the history of the plane. Worst of all, the logbooks are used to record compliance with airworthiness directives. At a minimum, you'll have to have an A&P verify or perform *every* AD that's ever been published on the airplane.

As you can imagine, that would be a king-size hassle. It might be worth it, if the purchase price were low enough. Talk to a mechanic in advance, and get an idea of how much money and paperwork is going to be required.

Tally ho!

In the natural course of things, you'll find a hot prospect. You do not want to simply shell out the money, though. You will want to look it over thoroughly beforehand. Chapter 7 examines the aircraft checkover and the actual purchase.

CASE STUDY: THE COHN 'RANGER

David Cohn's dream airplane was in front of him, all along. He learned to fly on Cessnas and got a taildragger checkout in a local flying club's Champ, but that wasn't enough. He wanted his own plane, preferably an unusual one. The Champ was all right, but his biggest wish was to find a plane just like the plane flown by his friend, Tom.

Tom was a member of David's university flying club. Tom and his dad had restored a Commonwealth Skyranger, a bit of flotsam from the great airplane bust of the late 1940s. Formerly the Rearwin Company, Commonwealth had made a nifty little two-seater called the Model 175 before the war. It became the Model 185 Skyranger. They retained the welded-steel-tube fuselage and wooden wing and installed an 85-hp Continental. The company started production and waited for the orders to flood in. It never happened. Only 350 were built before the company folded.

Cohn liked a lot of things about Tom's Skyranger. It performed about the same as a Cessna 140, but the big and comfortable airplane attracted a lot of attention wherever it went. The 'Ranger is so rare, few people can identify it.

Cohn kept in occasional touch with Tom after leaving college and moving a few hundred miles away. One day they happened to meet at a fly-in, and the Skyranger had a "For Sale" sign on it!

Purchase of the Skyranger took the entire $12,000 that David had been saving for a down payment on a house. Fortunately, his girlfriend Devon (Fig. 6-9) fully approved of and encouraged the purchase. They became engaged in the Skyranger two days after David bought it!

Fig. 6-9. Devon Wiel with David Cohn's Skyranger. Cohn proposed marriage to Wiel in the Skyranger. She accepted.

Ownership experience

David's first year with the Skyranger was an unqualified success. He spent $600 on the annual, plus another $400 on miscellaneous maintenance items. An outside tiedown on a largish airport near Boston costs $85 per month. Insurance for liability and hull coverage runs another $900 per year.

An old, out-of-production aircraft like this can cause some parts nightmares, but the Continental C-85 is no problem, and most airframe parts are so simple that they can be fabricated by any good A&P. With David lacking any mechanical background or tools, finding a good mechanic for this old airplane was important. He had incredibly good luck. Just down the road was an A&P who not only had restored two Skyrangers, but retained a parts stockpile.

The relatively large airplane powered by an 85-hp engine doesn't get anywhere fast. By the time Cohn had arranged to buy the Skyranger, he had a new job on the opposite side of the continent. "It took two weeks of low-and-slow that included seven days of actual flying time, three days of waiting out the weather, and a few more just visiting friends along the way. Old friends and new: The 'Ranger makes friends along the way," Cohn says.

He's still making friends. The field that he's based out of is home to a number of corporate jets. The 'Ranger is a jet-pilot magnet. "They all want a ride, and reminisce about what they learned to fly in. I've been promised rides in a TBM-700, a Gulfstream, and an Army Blackhawk helicopter," Cohn said.

The jet jockeys learned to fly in taildraggers; Cohn transitioned to them after starting out in Cessnas. His Champ instructor gave him 10 hours of dual in his flying club's Champ before turning him loose to solo. "It was before I felt confident enough," he admits. "It probably took another 40 hours [in the Champ] before I really felt confident putting it anywhere. And then another 20 before I actually *understood* the taildragger." All the practice held him in good stead with the Skyranger.

Like most converts to conventional gear, Cohn is now a diehard taildragger fan. One thing he misses from the Champ, surprisingly enough, is handpropping. "I remember everyone wanting to handprop (the Champ) when I went to other

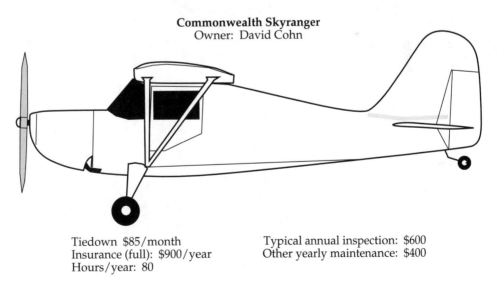

Commonwealth Skyranger
Owner: David Cohn

Tiedown $85/month
Insurance (full): $900/year
Hours/year: 80

Typical annual inspection: $600
Other yearly maintenance: $400

Fig. 6-10. A summary of the Skyranger's costs of ownership.

fields. Not that I didn't trust them, I just wanted the fun for myself!" The Skyranger, alas, has an electric starter. A summary of his costs of ownership is given in Fig. 6-10.

Advice

David Cohn is a great advocate of basing the aircraft nearby, even if it costs more. "It's nice to have the ability to just go out and check up on the plane. 'Oh, and now that I'm here, I'll just put on a few laps . . ..'"

He enjoys the rarity of his bird. "There's a lot of satisfaction in having people come out of the airport restaurant in the middle of their meals to ask questions about the 'Ranger. If I got a percentage on all the 'what kind of plane is that?' bets, I'd be able to make this plane pay for itself."

"Don't automatically assume you want a Cessna, Piper, or Mooney," he concludes. "Fly a bunch of different things before you buy. Speed? Ha! What's speed when you can *fly*!"

7

Found it!

YOU'VE FOUND THE PLANE OF YOUR DREAMS, you hope. Now you have to make sure that everything's in proper condition before dickering with the owner.

INITIAL CHECK

You probably aren't an A&P with an IA, but that should not prevent you as a pilot and potential buyer from being the first line of defense when considering the purchase of an airplane. You'll be able to find obvious problems that will immediately eliminate an airplane from further consideration. That will put you in a good position when you find everything is OK and you proceed to a prepurchase inspection. At $200–$1,000 you would like to pay for only one prepurchase inspection.

How do you start? Perform the most thorough preflight you've ever done in your life. Make note of the questionable points, and make sure to mention them to the A&P before she pulls her inspection.

First impressions

Start by standing 20 feet away and just looking at it (Fig. 7-1). Does the plane sit level? Is it perhaps cocked over, leaning to the side as if it's favoring one gear leg? For a fixed-gear Cessna, that's an immediate cause for worry. Cessnas have one-piece steel landing gear. If a leg is bent, there's been one heck of a slam.

Other airplane types might lean merely due to a soft oleo strut, but why is it soft? Is it leaking? Make a note to check the logs. If a mechanic has been fiddling with that side's oleo, perhaps it isn't fixed yet.

Then again, perhaps one wing tank is full and the other empty: check. Walk around the plane and examine the way the cowling and fairings hang. Are they a little crooked? Problems might be caused by some worn bolt holes or warped fiberglass.

Look at the paint to determine if some pieces are painted differently. They might be replacements from an aviation salvager. Nothing is really wrong with

Fig. 7-1. Step back from an airplane to take an overall look and make sure that everything appears normal before starting an up-close inspection.

that per se because the parts are usually quite airworthy, but you'll want to know why replacement was necessary.

The paint

Move in and look more closely. How's the paint? Ratty and patchy? Paint has two functions on an airplane. First, paint decorates the plane and makes it look pretty. Second, paint protects the underlying aluminum, wood, fabric, and the like, from the ravages of the environment.

We all prefer shiny new paint jobs; however, if cash is more important than cosmetics, you can find some pretty good deals on planes with lackluster paint, *as long as the airframe is protected.* (True, you'll have to get an A&P's opinion on that.) Poor paint on a fabric airplane can cause premature deterioration of the fabric covering material. The all-metal airplanes of the last 30 years do pretty good at standing up to the ravages of the weather. Nice paint doesn't make the plane fly any better.

Still, if a spouse has grudgingly given their approval to the aircraft purchase, buying something with grubby paint isn't likely to bolster their confidence, and shabby-looking planes are a bit embarrassing.

Exterior examination

A spiffy paint job might hide rot underneath. Take a good look at the paint. Any funny curlicue ridges that look like a worm track underneath the paint? That might indicate filiform corrosion. Is it flaking off anywhere? Are the steel parts rusty? Flaking and rust might indicate inadequate or nonexistent prepainting preparation, or just plain hard use (Fig. 7-2).

Looking farther, try to get a feel for the material underneath the paint. Ripples and dents in an all-metal airplane should trigger a "red alert." Wrinkles in fabric should raise a warning flag.

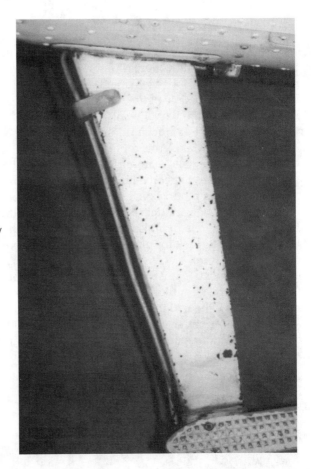

Fig. 7-2. The dark spots are flecks of rust. This 172 probably has a lot of time operating from a gravel runway.

Examine the control surfaces, the flaps, exposed control cables, pulleys, and the bolt and screw heads. Look for rust and missing pieces. Move the surfaces around, checking for free play. Listen for odd noises because a creak, scrape, or groan might indicate a frozen bearing in a pulley or cables that are too tight.

Look at any external fixed control tabs. These are pieces of aluminum attached to the control surfaces to minimize control pressure in level flight. Many planes have them on the rudder, a number of them have tabs on the elevator, and a few carry a fixed tab on an aileron. A fixed tab that is bent at a steep angle indicates that the aircraft is misrigged or out of true. One on the aileron typically means a serious wing heaviness problem because most airplanes have better means for correction.

Check the wheels and tires. Look for the red stains of brake fluid because bad brakes will cost more to fix than worn tires, and you can change the tires yourself.

While you're crouching, check the belly. Most have some oil and soot stains, but an excess of either is cause to wonder. If the paint looks a bit different on the belly of a retractable, there might have been a gear-up landing in the past. Not that the repairs would have just required a coat of paint, often the aluminum skin must be replaced, and that would show new paint.

Cabin and avionics

How's the cabin? Do the doors fit well? Are the seats well-upholstered and comfortable? Are they tight on their tracks? Any funny smells, like fuel or mildewed carpet? Does the panel have the avionics you want? This last point is important. Some people figure, "Well, I'll buy a cheaper plane with obsolete avionics and upgrade it to what I want."

Mention this to a bunch of aircraft owners and watch for those who cringe. They've probably taken this route. It's not just the expense; it's the delay and frustration. The plane will be grounded for quite a while, and you'll take it back to the shop several times to get the little things straightened out.

The hassle is proportional to the complexity of the work. Replacing an older navcom with a new model isn't all that bad, especially if you fly from an uncontrolled field where a radio isn't strictly required. But if you want a full-house up-to-date IFR suite, look for planes that are already equipped.

Check on two pieces of avionics. First, see if the plane has an encoder-equipped (Mode C) transponder. Vast areas still don't require it, but lacking Mode C will limit cross-country utility. Older planes like Cubs and Champs generally might not have them. If the plane was certified without an engine-driven electrical system, a transponder isn't required within the 30-nautical-mile veil under and around Class B airspace (Fig. 7-3). The plane can't *enter* Class B and C airspace,

Fig. 7-3. Because this Aeronca Chief doesn't have an electrical system, it doesn't need a transponder unless it actually enters Class B or C airspace. A simple hand-held 666 transceiver suffices for this owner.

though. The second piece of avionics is the emergency locator transmitter (ELT). For all intents and purposes, all planes are required to carry one; the major exception is single-seat aircraft. Check when the ELT's battery is due for replacement.

How are the instruments? Are they readable? Is their glass cracked? What about the trim molding around the panel and the glare shield? Glance around and check the Plexiglas. Is it yellowed? Cracked? Crazed?

Work the controls. Do they feel sloppy? Too much play can mean loose or worn cables, pulleys, bearings, and bushings. Listen for squeaks and scrapes. Run the trim system up and down, looking toward the tail to verify operation.

The engine

The engine is one area most of us will leave for the inspecting mechanic, but you can do a few things to weed out the also-rans. Ask the owner to remove as much of the cowling as practicable (Fig. 7-4). You can't judge the engine by peeking through the oil door.

Fig. 7-4. Some airplane cowlings open enough for an engine inspection, but others must be removed.

Again, you're performing an extremely thorough preflight. Look for cracked hoses, frayed wires, gouged ignition cables, broken baffles, and leaks (Fig. 7-5). Look for anything that just doesn't feel right: stains beneath cylinders, rubbed or

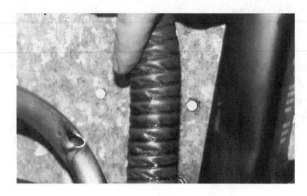

Fig. 7-5. Hole in the breather tube? Yes, but it's supposed to be there in case the end freezes up. It would be best to ask the mechanic during the prepurchase inspection just to be sure.

scraped spots, strong fuel odor, scraping or binding when the throttle or carb heat controls are worked.

Examine the exhaust system. Cracked pipes, mufflers, or heat exchangers cause expensive repairs. Look at the end of the exhaust pipe. Grayish or tan ash is normal for aircraft burning aviation fuel; it's a residue from the lead in the gasoline. Autogas leaves a black soot deposit. Oily or greasy deposits are something to ask the mechanic about.

Open the battery box. Any signs of stains around the fill caps? Perhaps the regulator isn't doing its job, and the battery is boiling over. Any staining or deterioration around the inside of the box? Maybe the battery is leaking acid, which is bad news indeed if it's been dribbling on aluminum.

Don't forget the propeller. Any sharp nicks? Do both tips match in shape? If it's a constant-speed model, are the blades tight and does the hub show an oil leak? If it's wood, is the varnish still shiny? Is the leading edge protection solid and whole? How's the spinner? Check for cracks and missing screws or bolts.

TEST FLIGHT

Now comes the fun part: flying. Your main purpose in the test flight is to determine that everything works the way it should. Have the owner start the engine. Watch if any weird little quirks are used. If the owner madly pumps the throttle during the start sequence, it might mean the carburetor accelerator pump isn't working right. Or it could mean that the owner doesn't really understand the engine and might have caused excessive wear through ignorance.

Leave the window or door open during start. Listen to the engine. Does it sound normal? Perform a careful runup. Does it run fine on either magneto? Remember that engines that are properly set up *will* show a mag drop. Too often a proud owner will say, "Look, only a 10-rpm drop on the left mag." In reality, the magneto timing is probably off. If there is no drop at all, the magneto switch could be faulty.

If the engine runs rough during the mag check or the rpm drops excessively, a number of things might be wrong. Express your regrets to the owner and taxi back to the ramp. You don't want to buy it with such a question mark. You don't even want to fly it. Let it get fixed out of the current owner's pocket, not yours.

The owner might not let you perform the takeoff or landing. Who can blame them? They know absolutely nothing about your skill level. They don't know how you'd handle a takeoff emergency, or even if you have enough hours to keep the plane under control. (When I sold my 150, I turned the controls over to one prospective buyer when we reached pattern altitude. He showed an inability to even fly straight and level. I suspect he'd seen the advertisement and figured it was a cheap way to take his first airplane ride.)

While the owner performs the takeoff, you can monitor the aircraft's performance. Watch the engine gauges during takeoff. Does the engine come up smoothly to full power? For a fixed-pitch propeller, the rpm should be around 80 percent of its redline at the beginning of the takeoff roll. From that point, it should increase gradually as airspeed rises.

Get a feel for the rest of the aircraft, too. Does it seem to be weaving around a bit? It might be a crosswind, but maybe the gear is a little out of alignment. The usual cause is a hard landing. Be alert for any shaking up front on a trigear airplane. One of their banes is a tendency for nosewheel shimmy.

When the plane breaks ground, check the rate of climb. Considering the density altitude, does it seem to match the book? Remember that the book values were determined by expert test pilots on brand-new airplanes. However, if published figures claim an 800-fpm rate of climb and the gauge only shows 400, there's cause for question.

The owner should let you take over at altitude. You don't have to perform the whole private pilot test repertoire; just do turns at various roll rates and banks (Fig. 7-6). How does the airplane feel? Are the controls tight and smooth, or sloppy and lifeless?

Set cruise power, trim up the elevator and release pressure on the controls, including the rudder. Does the airplane track straight? Most planes will eventually drop a wing slightly and start a gradual turn. Especially watch the slip-skid indicator. Misrigging in yaw can be irritating and cause misbehavior during stalls.

Verify that everything works, especially the avionics. Check the VOR against a local station. See if the loran or GPS gives your proper position. Fly an ILS approach if the plane is so equipped. Tune the radio. Are other aircraft and FAA facilities coming in loud and clear? Cycle the flaps, landing gear, and propeller, monitoring for proper operation and improper noises.

Above all, monitor yourself. How do you feel in this plane? Are you happy with the way it flies? Could you spend a week flying cross-country without yielding to the urge to take a fire-ax to the interior?

Head back to the airport monitoring all aircraft components and asking yourself crucial questions that come to mind. If everything is OK so far, it is time to review the logbooks.

CHECKING THE RECORDS

Logbooks are fascinating. I used to take my 150's books out once in a while and just scan through them. You can see where the plane's been and what's been done with it. You can touch the signatures of the previous owners and wonder what

Fig. 7-6. The proof is in the flying. A test flight can tell you a lot about the condition of the plane's systems.

they're flying now. Just from the number of entries and their thoroughness, you can tell who took care of the plane and who didn't.

Examining the logs of an unfamiliar airplane is important to get a feel for the condition of the aircraft, but the logs don't tell you anything right out. It takes a bit of detective work and inference to extract the true background of the airplane.

Take when I bought my 150. The sign in the window said, "Never a Trainer." I was skeptical. I didn't believe that a 20-year-old Cessna 150 could have escaped being flogged around the pattern by students at some point in its life. Sure, the use of the aircraft isn't indicated in the logs. But there were entries about annuals being performed by "XYZ Flying Club." There were periods when the plane flew up to 300 hours between annuals. That wasn't just a happy owner flying around on the weekends.

In addition to such routine entries, the FAA requires that any repairs or inspection also be noted in the logbook. Unfortunately, the FAA doesn't require that the *cause* of the repair be mentioned. It's in your court to read between the lines.

This entry in my 150's logs set flags waving: "Replaced leading edge skin between outboard ribs." Why would they do that? Damage, obviously. And damage on the leading edge generally means the airplane hit something. Probably on the ground, since midairs generally negate the need for further log entries.

Was the leading edge damage significant? I didn't think so. It had happened 15 years before. If it had been more recent, I would have worried about damage that the repair mechanic might have missed. Hitting a pole with the wingtip can cause more than external damage; the leverage can tear metal in the wing's root and even the fuselage. In my case, the plane had undergone 15 annual inspections since the repair. I figured that at least one of the IAs would have found any additional damage.

Don't forget the engine and propeller logs, as well. They can tell their own stories of problems and damage. If you find an entry in the engine log regarding checking the "crankshaft runout," beware. This is typically done after a propeller strike, the ramifications of which were discussed in chapter 2. A crankshaft runout determines if the shaft was bent by the propeller impact.

By the way, not all aircraft have a separate log for the propeller. For a long time, planes with fixed-pitch props kept a combined "Engine and Propeller" book. The FAA now encourages separate logs, even for fixed-pitch props.

Engine cylinder compression is checked at each annual; the results are required to be entered in the engine log. The test procedure isn't at all like that used for cars. You'll see a set of numbers such as "80/72" (they might be the opposite way "72/80") listed for each cylinder. Chapter 13 examines what these numbers mean. For now, you should know that if the smaller number is below 60, the engine will probably require cylinder, ring, or valve work.

Track the compression values over a couple of annuals. If one cylinder's compression drops suddenly, beware. There are a few legitimate, noncritical reasons for a temporary reduction, but you might end up buying a new cylinder all too quick.

Every replaced aircraft component should be accompanied by paperwork to prove that it was fully FAA-approved and airworthy. If the engine log shows that a magneto was replaced, the *yellow tag* that came with the magneto should have been retained. Look for these tags. Don't forget to check on the ADs for the airframe, engine, propeller, and appliances, too.

Sometimes putting all three logbooks together can be enlightening. Say you're looking at a 1972 Mooney. You notice that the propeller log starts in May 1980 with the installation of a new propeller. Why was it replaced? The engine log shows that the engine was torn down and inspected at about the same time, and the airframe log shows the belly skin was replaced as well. It should be very apparent that the plane landed gear-up in early 1980.

Does it make a difference? That far back, probably not. But if the owner is claiming "no damage history," you've got leverage when you start dickering over price. Don't immediately suspect the owner of trying to pull a fast one. He might not have owned it at the time, and he might not have checked the logs carefully when he bought the plane.

The logbooks will also tell you about the modifications that have been performed the aircraft. If the owner tells you that the plane is approved for auto-gas, you should be able to find an entry recording execution of the appropriate STC. There should also be a copy of the FAA Form 337 that was executed and sent to the nearest flight standards district office. All modifications require the filing of a Form 337; if the aircraft has been modified, but the paperwork hasn't been filed, the aircraft isn't legally airworthy. If you buy an airplane with insufficient modification paperwork, you will have to pay a mechanic to check the mods and file the appropriate paperwork.

If you want to verify that the paperwork has been filed and accepted by the FAA, you can request a copy of that aircraft's records. Write:

U.S. Department of Transportation
Federal Aviation Administration
Mike Monroney Aeronautical Center
P.O. Box 25082
FAA Aircraft Registry, AAC-250
Oklahoma City, OK 73125

The search costs $2, the first page of the records costs 25 cents, and each additional page costs 5 cents. If the records are on microfiche, there's an additional 15-cent charge. You can call them at 405-686-2116, but people tell me that it's almost impossible to get through.

Finally, as you go through the logbooks, determine the date of the last annual inspection. If the annual is due next month, that's another $500–$????? that'll be coming out of your pocket almost immediately. There are other required inspections as well. The transponder components must be checked every 24 months, and a static system test is required at the same interval to be legal for IFR operations.

Again, you don't have to be all-knowledgeable about aircraft repair. Pick out sections that you find unusual, and talk to an A&P.

Engine matters

As I've mentioned earlier, the condition of the engine is one of the major price drivers, especially for smaller, simpler planes. If your prospective purchase has a "high-time" engine, it's good to be cautious. Expect to pay less.

Just remember, if the manufacturer lists an 1,800-hour TBO and the engine has 1,700 hours SMOH, that doesn't mean you'll get only a hundred hours of flying out of it. I have a friend who is rebuilding a Piper Tripacer. The engine is 400 hours past the TBO, yet he isn't going to rebuild it. The Lycoming runs strong and has shown no signs of distress.

How an engine has been operated is more important to its longevity than *how long* it has been operating. The Lycoming in a 152 is going to perform a lot of touch-and-goes, which translates into a bunch of full-power climbouts immediately followed by idling glides. It's under a lot more stress than an engine used in a fish-spotting plane that spends almost all its time at cruise rpm. They've been known to go as far as *1,000* hours past the recommended TBOs.

Still, TBO-busting is a nerve-wracking process. A good number of perfectly good engines have been overhauled just because the owner got jittery as the tachometer's hour meter neared the magic number. But even if the engine in your prospective purchase has only a few hours since its last overhaul, you aren't home free. The quality of overhauls varies drastically.

Two local pilots bought a Cessna 140 with a Continental C-90 that had 300 hours since overhaul, which would be considered a low-time engine. Within a year they spent $3,000 on the engine because some of the internal parts were garbage. Those parts did not get trashed in 300 hours.

Any licensed A&P can overhaul an engine. How good the overhaul is depends on how finicky and (to be blunt about it) honest that mechanic is. The procedure seems simple enough. The overhaul manual for the engine specifies the maximum amount of wear a part can show and still be used on the engine.

Yet there are ways around that to stretch the specs enough to get marginal parts to pass. It might be done as a favor to a friend, to give the engine an "overhaul" at minimum cost. It might be done to maximize profit. The person performing the overhaul might have been lazy or sloppy.

So if an engine is listed as being recently overhauled, dig a little bit and try to determine the quality of the overhaul. If it was performed by the factory or one of the nationally known engine shops, so much the better. Always remember that local mechanics can perform an overhaul just as well. Be prepared to ask the A&P performing the inspection about the overhauler. Ask the owner for the receipts from the overhaul to help determine what work was done.

There's one way to positively ensure that the overhaul is done to your standards. Buy a plane with a broken or runout (past TBO) engine, then have the engine rebuilt at a quality shop. It'll require a greater initial cash outlay beyond the purchase price, but you'll end up with an engine that you can trust.

THE PREPURCHASE INSPECTION

Looking the plane over yourself is a great start, but with the kind of money at stake, a prepurchase inspection is a very, very good idea. Recall the second case study in chapter 4. Mike Furlong bought a ramp-queen Stinson after a talk with the owner and a cursory look-over. It took him nine years and another $20,000 before he flew his new plane.

It doesn't always happen that way. At the time that I bought my 150, I hadn't flown in seven years. I'd never flown in the state in which I was then living. I had no contacts, nor knew where to find an A&P. I read the logs, gritted my teeth, and handed over a check. Everything turned out OK. But in retrospect, I did not proceed properly. I won't do it again.

How do you find a mechanic to do the prepurchase inspection? Start by looking in the aircraft logbooks to find who did the last annual. Scratch that mechanic off your list. That mechanic is probably competent, but if he missed something at the last annual, he just might miss it again. Or if he finds it, he might be too embarrassed to admit it.

If you have any friends who own airplanes, see if they'll recommend someone. Otherwise, just call some local FBOs. All shops are familiar with prepurchase inspections and should be able to give you quotes over the phone. If the plane is an unusual one, spend some time to locate a mechanic with experience in that model.

Depending upon the complexity of the aircraft, a prepurchase inspection will probably run from $150–$500. At the utmost minimum, you want the mechanic to give the engine a clean bill of health plus determine if the plane is legal:

- The ADs are up to date.
- All maintenance was performed and logged properly.

- No life-limited parts (some hoses, the ELT battery, etc.) have expired.
- All required markings and placards are in place.
- All inspections (annual, static system, transponder, and manufacturer-required accessory checks) are current.

That level of inspection will make sure you (probably) won't have to deadstick on the way home, and ensure that any flying you do is legal. There's a lot more that *should* be done, though. If your mechanic is experienced in the aircraft model, she should know where potential problem areas lie, where corrosion tends to appear, what tends to form cracks, what tends to break, and what tends to wear.

An airplane has few "trivial" problems. A broken window latch sounds minor, except that a new one might only be available through the factory and might cost $100 or more before you add the installation charge. You want to know as much as possible about the airplane's condition before you're the one responsible for it.

The prepurchase inspection is *not* the place to scrimp. The best prepurchase inspection equals an annual inspection without any cleaning, lubricating, or repairing. If the annual is due, you might be able to work a deal with the seller: You'll pay the *inspection cost* of the annual, if the seller pays for the *repair costs*.

Prepurchase inspections aren't covered by any regulation; hence, the results aren't noted in the aircraft logbooks. Instead, the mechanic should give you a summary of his assessment of the aircraft's condition. The mechanic should be able to give you an informal estimate about the costs to repair the deficiencies found.

WHO OWNS IT?

Before you contemplate handing a large check over to this aircraft-selling stranger, you'd better verify ownership of the airplane in question. Maybe the owner just didn't bother to file the required paperwork when the airplane was purchased, and the title is now in question. Perhaps the plane is collateral in a bank loan. Maybe it's stolen.

If you complete the transaction and it turns out that some third party actually had valid title on the plane, they get it—no ifs, ands, or buts. You're out the purchase cost, unless you can find the seller and extract the cash. Good luck.

The title search

Like a prepurchase inspection, verification that the seller has legal authority to sell you the airplane can save you a bundle later. You can find this out through a *title search*. Unlike a lot of financial/legal issues, understanding title searches is simplicity itself. The search identifies the current legal owner of the aircraft. If the name on the title differs from that of the person with a hand outstretched to take your check, run.

Unlike a prepurchase inspection, title searches are low-cost and painless. The price is generally fewer than $50. Results are usually received in a day. Get a title search, no matter who's selling the aircraft: family member, friend, FBO, broker, or national sales firm. If the seller tries to talk you out of a search, run. The penalty for error is just too severe. Get one. No exceptions.

Title search companies

Many aircraft publications include advertisements for companies that perform title searches. In *Trade-A-Plane*, for instance, look under "Aircraft Title Service/Legal Services." AC 8050-55, *Title Search Companies*, is available from the FAA. It lists those companies approved to perform title and record searches. The AOPA also performs this service. Their number is 800-654-4700.

Paying off the seller's bank

If the owner used the aircraft as collateral for a loan, the finance company will be listed as the owner of record. Buying the plane in this case will be similar to buying a used car where the seller still owes money. You'll have to actually pay the bank, who'll sign over the title and pass the cash left over from paying off the loan to the seller.

THE DEAL

Did it pass the tests? If so, we're on the home stretch.

Dickering

Ever bought a used car? The negotiation process for a used airplane works similarly. The owner claims the plane is a rarity in superb condition, and you'll argue that it's a piece of junk that you might be willing to take off their hands.

Actually, it's best not to go quite that far, at least with aircraft. If the plane meets your requirements, you don't want to antagonize the owner. Small egos are rare in aviation. Tick off the owner, and he might decide not to sell to you at any price.

And what *should* that price be? You've already done your research and know what equivalent aircraft are selling for. Your mechanic has identified several items that'll need repair, so deduct the cost from the estimated worth.

Base your dickering with the owner on the aircraft's estimated worth. It goes without saying that you'll want to offer less. There's nothing worse than making an offer that the seller accepts with lip-smacking relish.

Have your research materials available. Be ready to show the owner some advertisements for equivalent airplanes; point out the flaws your mechanic found. As cold-blooded as it might sound, your best hope for saving money lies in an owner who didn't do any research on what the plane is worth.

Which makes dealing with brokers or FBOs that much less satisfying. They'll not only try to hold out for top dollar, but their commissions or markups run the price even higher. Overpaying for anything, including airplanes, leaves a bad taste in one's mouth. You'll make a lot better deal with a private seller.

With a private individual or aircraft broker, you'll sometimes come to an impasse. The seller has come down as far as he or she will, and it's thousands above what you want to pay. Your response can take a number of forms. The "book" response here, of course, is to walk away. You know what the airplane's worth and shouldn't pay a penny more.

Unfortunately, in the real world, things aren't that easy. Sometimes you really *want* that particular airplane. Maybe it's a rare model that you've looked high and low for. Maybe it's equipped exactly the way you want. Should you bite the bullet and pay the asking price?

It's impossible to make that decision in advance. Hopefully you know the maximum that you can afford. And no matter how good the airplane is, it'll probably cost more to own than you expect. Don't put yourself even farther in the hole by paying more than is comfortable.

There are a couple of alternatives. First, you can politely take your leave, giving your name and phone number to the owner. If a month or so goes by without the plane selling, they might be more interested in your offer. Second, you can offer to pay the asking price, but make the deal contingent on associated issues. For instance, you might suggest the owner pay for a fresh annual inspection. Even if the plane had been inspected just six months earlier, a new annual would push the date of the first one that you pay for that much farther back. If the plane's being sold through an FBO, they can often get the annuals at a discount rate.

Other issues are possible with a private owner. Perhaps you're three years down on the hangar waiting list, and the seller already has a hangar at your home field. Offer to pay the higher price if the seller subleases the hangar to you for the next two years (if the hangar lease and airport rules permit). Maybe the seller of a rare plane has a stock of spare parts; include the spare parts with the airplane if you can't get the selling price down.

The handover

Hopefully, eventually, you'll agree on price. The next thing to work out is how to pay. Would you accept someone's personal check for $30,000? Especially when they're going to take possession of valuable and highly mobile property?

I bought my 150 with a certified check; the cost is minimal and the seller's risk low. When I sold the plane, the buyer paid cash! A veritable pile of $100 bills. I hadn't known he was going to do that. We'd agreed for the handover at 4:30 on a Friday afternoon. At 4:40 I had a bulging envelope full of freshly counted bills and a sinking feeling that I'd be staying up all weekend standing guard over it. (I get nervous when I have a $50 bill in my wallet.) Fortunately, I arrived at a nearby bank before it closed.

Not all purchases are this simple. Sometimes the seller wants to see a little "earnest money" up front. Or the seller promises to have the aircraft annualed if you agree to buy it. Or the seller lives in one location, you live in another, and the bank that holds the lien on the airplane is located far away from both parties.

Obviously, handling the paperwork and money transfers in these cases is awkward. An escrow agency can come to the rescue. They act as a disinterested third party, ensuring that the money and title are not transferred unless agreed-on conditions are met and that all required documentation is properly submitted. Call a couple of agencies and talk about your upcoming transaction. AOPA's title and escrow service can be reached at 800-654-4700.

If your transaction is simple and you tackle it yourself, make sure the plane has the following before you hand over the lucre:

- Airworthiness certificate (FAA Form 8100-2)
- Weight and balance/equipment list
- Flight manual or operating limitations
- Airframe log
- Engine log
- Propeller log (or combined engine-propeller log)

(Sharp-eyed readers will note the omission of two items that an aircraft normally requires: registration and radio license. The start of chapter 8 will explain.)

When everything is met to your satisfaction, give the owner the money and have him or her fill out and sign the FAA Aircraft Bill of Sale, AC Form 8050-2 (Fig. 7-7). Congratulations. You have an airplane! Now comes the hard part: actual ownership. The rest of the book covers the legal, mechanical, and emotional aspects of your new possession.

Other sources

Buying an airplane is a nerve-wracking, complex operation. You won't hurt my feelings a bit if you decide to read what other authors have to say about the process. Two other TAB/McGraw-Hill books cover aircraft buying: *The Aviation Consumer Used Aircraft Guide* and *The Illustrated Buyer's Guide to Used Airplanes* by Bill Clarke.

Don't forget books specializing in particular aircraft or aircraft lines. In addition, a type club might be able to provide helpful hints. If you're an AOPA member, the organization has a nifty little packet with tons of information and advice, plus copies of all the required FAA and FCC forms. Call 800-872-2672 and select the member services department.

CASE STUDY: A NEW PIPER—FREE!

Shopping for and choosing an airplane can be a hassle. There's a perfect way to avoid it: win a brand-new airplane. It's a solution not without its own unique problems. Sure, it would be nice to have a clean new flying machine, but successful ownership is still going to take a lot of money and work. If you don't have much of the former, you'll need more of the latter.

Margaret Puckette (Fig. 7-8) didn't plan on winning an airplane. All she did was buy one. An artist and art instructor in Corvallis, Oregon, she caught the fly-

UNITED STATES OF AMERICA
U.S. DEPARTMENT OF TRANSPORTATION FEDERAL AVIATION ADMINISTRATION

AIRCRAFT BILL OF SALE

FORM APPROVED
OMB NO. 2120-0042

FOR AND IN CONSIDERATION OF $ **15,900** THE
UNDERSIGNED OWNER(S) OF THE FULL LEGAL
AND BENEFICIAL TITLE OF THE AIRCRAFT DES-
CRIBED AS FOLLOWS:

UNITED STATES
REGISTRATION NUMBER **N 1234**

AIRCRAFT MANUFACTURER & MODEL **Cessna 172 B**

AIRCRAFT SERIAL No. **172 5557**

DOES THIS **23d** DAY OF **March** 19**95**
HEREBY SELL, GRANT, TRANSFER AND
DELIVER ALL RIGHTS, TITLE, AND INTERESTS
IN AND TO SUCH AIRCRAFT UNTO:

Do Not Write In This Block
FOR FAA USE ONLY

PURCHASER

NAME AND ADDRESS
(IF INDIVIDUAL(S), GIVE LAST NAME, FIRST NAME, AND MIDDLE INITIAL.)

**Loft, Ima L.
1515 Flight Road
Anycity, Tx 22222**

DEALER CERTIFICATE NUMBER

AND TO   EXECUTORS, ADMINISTRATORS, AND ASSIGNS TO HAVE AND TO HOLD
SINGULARLY THE SAID AIRCRAFT FOREVER, AND WARRANTS THE TITLE THEREOF.

IN TESTIMONY WHEREOF   HAVE SET   HAND AND SEAL THIS   DAY OF   19

SELLER

NAME (S) OF SELLER (TYPED OR PRINTED)	SIGNATURE (S) (IN INK) (IF EXECUTED FOR CO-OWNERSHIP, ALL MUST SIGN.)	TITLE (TYPED OR PRINTED)
M. Grounded		Seller

ACKNOWLEDGMENT (NOT REQUIRED FOR PURPOSES OF FAA RECORDING; HOWEVER, MAY BE REQUIRED
BY LOCAL LAW FOR VALIDITY OF THE INSTRUMENT.)

ORIGINAL: TO FAA

AC Form 8050-2 (9/92) (NSN 0052-00-629-0003) Supersedes Previous Edition

Fig. 7-7. Bill of sale form.

Fig. 7-8. Margaret Puckette.

ing bug from several friends. She started ground school in early 1989. She decided she needed a plane to take lessons in and bought a 1977 Grumman AA-1B Trainer.

At that point, she hadn't even taken her first flying lesson. "When I bought the Grumman," she remembers, "All I knew how to do was open and close the canopy." She couldn't even start the engine. Was this a recipe for disaster, in the form of a "skygoing" lemon? Not in this case because she knew Robert Grove, the Grumman's owner. He'd upgraded to a Grumman Tiger and needed to sell his beloved little Trainer. He'd taken excellent care of it, and Margaret was willing to trust him. "I knew, for a first-time buyer, I'd have help from the previous owner. I somehow knew I couldn't lose."

It worked out well. "I felt so high after my first lesson," she says. The Grumman was faultless as she worked her way toward her private license; however, there was one little side trip. Just after buying the plane, a friend pointed out that Margaret was now qualified for membership in the Aircraft Owners and Pilots Association (AOPA). She didn't even have a student certificate, but she was an aircraft owner. Convinced of the benefits of membership, she joined.

That year was the 50th anniversary of AOPA's founding. To celebrate, they were giving away a brand-new Piper PA-28-181 Archer. Signing up for membership entered her in the contest. In October, Puckette passed her private pilot flight test. Eight days later, she won the Archer. One small problem: The Piper was in Orlando, Florida, a continent away from her home in Oregon. The flight home would be quite a cross-country for someone with a new pilot's license.

AOPA, though, knew her background. Don Koranda, AOPA's senior vice president of member services, gave her a thorough checkout. Her training in the nimble Grumman paid off. It required better skills than the garden-variety Cessnas, so she was already a more-precise pilot.

Margaret chuckles over one aspect of the checkout. "Don was enthusiastic about how light on the controls the Archer was, but it was heavy, compared to my Grumman!" Her cross-country trek took three days and 28 flying hours. For safety's sake, a friend with a commercial license rode with her most of the way.

On arrival in Oregon, she and her husband faced the first big hurdle: Winnings, whether cash or property, are taxable. AOPA's accountants had sharpened their pencils to reduce the IRS's bite. They did a pretty-good job, but the Puckettes faced a $28,000 tax bill.

They managed to pay it without taking out a loan. She sold the Grumman. Their own accountant found some overpayments in previous years' tax returns, and Margaret received a commission for a painting.

Winning the airplane really stoked Margaret's flying bug. Archer N1939G (Fig. 7-9) was loaded, making it a perfect instrument trainer. She got her IFR ticket, then a commercial certificate, then a flight instructor certificate, then an instrument instructor rating.

Fig. 7-9. Archer Three-Niner Golf came fully equipped and was the springboard to the proud new owner's advanced certificates.

In the meantime, Puckette and her husband divorced. He got the house, and she got the airplane plus custody of their two preteen girls.

Most of us find airplane ownership a stretch. It's especially true for Margaret Puckette. She's an art instructor at a local community college. She derives some income from the sale of her paintings, drawings, and musical instruments. She has to balance the demands of single parenthood and those of aircraft ownership with an income that varies with the whims of the art market.

She could sell the Archer, of course. Even today, it's worth almost $100,000. She could buy an equivalent used Piper for half that, and live a bit more comfortably, but Archer N1939G is now a member of the family. It's a part of her life and her identity. She's determined to keep it.

Ownership experience

With her variable income, Puckette, by necessity, works on the Archer as much as possible. "The most I'd worked on before was a bicycle. Now I do all the maintenance FAR 43 allows, up to and including replacing seat belts."

The annuals are owner-assisted, often educational for the teacher. "I have a fabulous mechanic. He likes to explain things as he works," she says. Involvement has paid off because the annuals run about $400, and the lowest was about $250.

The owner of a new aircraft doesn't have to contend with the years of wear and tear most owners face. To keep things running smoothly, Margaret started treating it right from the beginning, on the ground and in the air.

However, new planes have their disadvantages. She's handled a number of minor manufacturing defects. These include a flawed battery case and numerous screws and metal shavings rolling around underneath the floorboards near the control cables. She even found a footprint on the inside of the fuselage skins. One flaw might have contributed to a problem with the autopilot. (I *said* it was loaded!) The avionics cooling-fan hoses weren't quite attached properly, which might have caused some localized overheating.

Also, one of Archer N1939G's heater ducts is routed too close to the exhaust manifold. The manifold's heat makes the duct brittle, forcing duct replacement at each annual. The oil pressure gauge used to have a hard copper line to the engine. The line passed engine vibration direct to the gauge and caused its early failure. She replaced the copper with a flexible connection.

One AD required the addition of stall strips to each wing. Fortunately, this came out just after she took possession and was covered by the new-aircraft warranty.

She added a Northstar loran. The manufacturer sold her the unit at a special price because it knew the AOPA Archer gathered a lot of attention and figured it would be good advertising. In the same vein, Sigtronics donated an intercom.

Because late-model aircraft are so valuable, her insurance premiums are high. With the aircraft kept in a locked hangar and Margaret's flight instructor certificate with instrument instructor rating, her premiums are about $1,200 a year. The premiums were a bit higher back when she was a newly minted private pilot with a brand-new airplane.

Insurance came in handy, though. Just three months after she won the Archer, a windstorm blew the hangar door down onto the plane. The door bashed the wing, dented the spinner, and crushed both ailerons and a flap. One advantage to a new airplane is that you don't have to scrounge up parts from the aircraft junkyard. Archer N1939G received factory-new replacement parts. A local shop matched the AOPA Piper's custom paint job. Puckette's insurer covered the work without murmur.

Probably the most amazing thing about the Archer is how much use she's derived from it. The airplane is not used for instruction, except for her children. She flies it about 200 hours per year.

The Archer has had a significant effect on her art. Even before she started to fly, her paintings had an aerial perspective, imaged as if the viewer were 8 feet above the floor. The Piper has added a new element. Sometimes this new perspective is obvious, as seen in "A Flight Over Crater Lake" (Fig. 7-10). The painting not only features an aerial view of the lake, but adds the Piper's instrument panel in the foreground. Sometimes the new perspective is indirect, such as "Quiet Hangar." Lightplanes parked in a darkened hangar evoke the warmth and satisfaction at the end of a perfect day's flying.

Fig. 7-10. Puckette's painting, "A Flight Over Crater Lake," includes the Archer's instrument panel, not to mention her own left hand on the control yoke.

The Archer's effect is mostly subtle. Puckette's landscapes have a much higher point of view that is fascinating to the general public and instantly recognized by pilots.

She also volunteers her time and the Archer for an environmental group. Operation Lighthawk provides aerial platforms for those who wish to view the actual sites under debate. Since joining, Margaret has flown radio, newspaper, and TV network personalities, congressional and senatorial aides and staff members, attorneys, scientists, and media members from around the world. "I'm proud of the work I've done with Lighthawk," she says. "I've met some extraordinary people on both sides of environmental issues." Her work has been effective, too. She was recognized as Operation Lighthawk's "Conservation Pilot of the Year" in May 1994.

Archer N1939G's life isn't all work, though. She uses it for travel and vacation, not to mention socially. "It's a big plus to offer a flight to someone, but it's not just a fun date. I've flown to San Francisco for shopping." The Piper's AOPA paint job makes sure she gets recognized everywhere she goes. "You won my airplane," is the usual jocular comment. The typical costs of operating her aircraft are shown in Fig. 7-11.

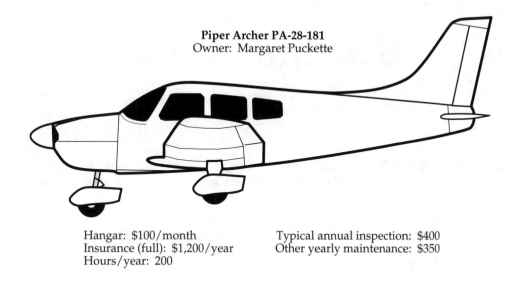

Piper Archer PA-28-181
Owner: Margaret Puckette

Hangar: $100/month
Insurance (full): $1,200/year
Hours/year: 200

Typical annual inspection: $400
Other yearly maintenance: $350

Aircraft won in AOPA contest

Fig. 7-11. A summary of Puckette's costs of ownership.

Her involvement in a male-dominated activity has one rather weird side effect. "I've had a lot of guys ask me to try and inspire their wife or girlfriend to get interested in flying. [With] something as major as an airplane, the partner *should* be supportive."

Advice

While she didn't actually pick out the plane she ended up with, Puckette has one piece of advice for prospective buyers: "Know the type of flying you want to do: long distance, cargo, fun, family, and the like. Find an airplane that suits your uses. Fit it into your life." Airplanes and flying require attention, and you can't just wall them off from the rest of your world.

She's a strong proponent of owner involvement in maintenance. Puckette said, "I don't understand people who won't put a screwdriver to work. You can get most of the simple tools you need at places like Sears or auto-parts stores. Get your hands dirty!" One added advantage of owner maintenance is that it adds to pilot confidence during preflight. With her continual involvement, Puckette can quickly see if something is amiss. Her final thought: "Get an instrument rating. It pays."

8

The first week

YOU WILL HAVE TO PLACATE three government entities now that you have an airplane: FAA, Federal Communications Commission (FCC), and your state's aircraft department.

COMPLETING THE PAPERWORK

The FAA

The previous owner filled out the AC Form 8050-2, the Aircraft Bill of Sale. Notice that the bill of sale is a two-part form; the FAA gets one part, and you get to keep the other. Don't lose your copy; other than the obvious legal issue, you might need to present it to local or state agencies, which is a subsequent topic of this chapter.

The other bit of FAA paperwork is AC Form 8050-1, Aircraft Registration Application (Fig. 8-1). The seller should have the old registration because he or she will need it to notify the FAA that they've sold the plane. The pink copy included with Form 8050-1 is a temporary registration; immediately place it in the aircraft. You're covered for the 90 days or so that it'll take for the FAA to send you a permanent registration. Because the registration application is also used for address changes, pick up an extra or two.

Note that the temporary registrations states "OPERATIONS OUTSIDE THE UNITED STATES ARE PROHIBITED BY LAW." Simple enough: Don't fly the airplane to another country, such as Canada or Mexico, until the permanent registration is on hand.

To formalize the transfer of ownership, send the white copy of the bill of sale, the white and green copies of the Registration Application, and a check for $5 (made out to "Treasurer of the United States") to:

Federal Aviation Administration
Aircraft Registry
AAC-250
P.O. Box 25504
Oklahoma City, OK 73125

UNITED STATES OF AMERICA DEPARTMENT OF TRANSPORTATION
FEDERAL AVIATION ADMINISTRATION-MIKE MONRONEY AERONAUTICAL CENTER
AIRCRAFT REGISTRATION APPLICATION

CERT. ISSUE DATE

UNITED STATES REGISTRATION NUMBER	N 1234
AIRCRAFT MANUFACTURER & MODEL	Cessna 172 B
AIRCRAFT SERIAL No.	1725557

FOR FAA USE ONLY

TYPE OF REGISTRATION (Check one box)

☑ 1. Individual ☐ 2. Partnership ☐ 3. Corporation ☐ 4. Co-owner ☐ 5. Gov't. ☐ 8. Non-Citizen Corporation

NAME OF APPLICANT (Person(s) shown on evidence of ownership. If individual, give last name, first name, and middle initial.)

Loft, Ima L.

TELEPHONE NUMBER: (101) 555-7777

ADDRESS (Permanent mailing address for first applicant listed.)

Number and street: 1515 Flight Road

Rural Route: P.O. Box:

CITY	STATE	ZIP CODE
Anycity	TX	22222

☐ **CHECK HERE IF YOU ARE ONLY REPORTING A CHANGE OF ADDRESS**

ATTENTION! Read the following statement before signing this application. This portion MUST be completed.

A false or dishonest answer to any question in this application may be grounds for punishment by fine and / or imprisonment (U.S. Code, Title 18, Sec. 1001).

CERTIFICATION

I/WE CERTIFY:

(1) That the above aircraft is owned by the undersigned applicant, who is a citizen (including corporations) of the United States.

(For voting trust, give name of trustee: _____), or:

CHECK ONE AS APPROPRIATE:

a. ☐ A resident alien, with alien registration (Form 1-151 or Form 1-551) No. _____

b. ☐ A non-citizen corporation organized and doing business under the laws of (state) _____ and said aircraft is based and primarily used in the United States. Records or flight hours are available for inspection at _____

(2) That the aircraft is not registered under the laws of any foreign country; and

(3) That legal evidence of ownership is attached or has been filed with the Federal Aviation Administration.

NOTE: If executed for co-ownership all applicants must sign. Use reverse side if necessary.

TYPE OR PRINT NAME BELOW SIGNATURE

EACH PART OF THIS APPLICATION MUST BE SIGNED IN INK.

SIGNATURE	TITLE	DATE
Ima Loft	Owner	3/23/95
SIGNATURE	TITLE	DATE
SIGNATURE	TITLE	DATE

NOTE   Pending receipt of the Certificate of Aircraft Registration, the aircraft may be operated for a period not in excess of 90 days, during which time the PINK copy of this application must be carried in the aircraft.

AC Form 8050-1 (12/90) (0052-00-628-9007) Supersedes Previous Edition

Fig. 8-1. Registration application.

The FCC

If your acquisition has a radio, you'll need to apply for a radio station license. FCC Form 404 (Fig. 8-2) is used; most FAA offices stock it. While the FAA forms are surprisingly understandable (for government paperwork), the FCC application is a bit more obtuse. For most privately-owned aircraft, enter "PAA" for item 8 and "0001" for item 9. As of this writing, the "fee due" (item 10) is $115.

The station license is good for any transmitters within the aircraft. You don't need a separate license for your backup communications (handheld). Mail the form and the check for $115 payable to the FCC to:

Federal Communications Commission
Aviation Aircraft Service
P.O. Box 358280
Pittsburgh, PA 15251-5280

The FCC form comes attached to a page of instructions. The back of the instructions includes a temporary license. Fill this out and stick it in the airplane. It's good for 90 days, just like the FAA's temporary registration. The new license is good for 10 years.

It's important to fill the FCC application out correctly. The FCC at one time proposed that it would automatically reject all erroneous applications and keep the money. Pocketing $115 for a typo seems rather steep. A number of organizations convinced the FCC that the proposal was not a good move, and the idea was scrapped. But Hell hath no fury like a bureaucrat denied revenue, so don't tempt fate. Double-check your application.

Even if your new airplane doesn't have a radio, apply for a license. You never can tell, you might end up borrowing a handheld someday to fly into a controlled field (Fig. 8-3). It would be just your luck to get hit with an FAA ramp check that day.

State authorities

About 40 states require registration of private aircraft, and 84 percent levy a sales or use tax on aircraft purchases. State registration fees vary; in my home state, they start at $50 for simple two-seaters and go on up.

If you buy a new aircraft, you'll pay sales tax (Fig. 8-4). If you buy it outside of your home state, the state revenuers will hit you for it when you register the aircraft with the state aviation department. In some cases, the FAA passes the information from AC Form 8050-1 to them, and they'll notify you of the need to pay a tax. If you buy the plane out of state, you should be able to pay just your home state's taxes. You should not have to pay both states' entire tax. In any case, get itemized receipts. Sometimes your own tax liability will be decreased by the amount you paid in another state.

In most states, you don't have to pay "sales tax" if it's a private-party transaction; however, you can be hit with a "use tax" that (surprise, surprise) is about the same percentage as the sales tax that you don't have to pay. Depending upon location, you'll have to pay 3–8 percent of the selling price of the aircraft, as shown on the bill of sale.

Approved by OMB
3060-0040
Expires 7/31/94
See instructions for
public burden estimate.

FOR
FCC
USE
ONLY

UNITED STATES OF AMERICA
FEDERAL COMMUNICATIONS COMMISSION

APPLICATION FOR AIRCRAFT RADIO STATION LICENSE

1. APPLICANT NAME	*Loft, Ima L.*

2. MAILING ADDRESS (Line 1)	*1515 Flight Road*

3. MAILING ADDRESS (Line 2)	

4. CITY	*Anycity*

5. STATE *TX*	6. ZIP CODE *22222*	7. FAA REGISTRATION OR FCC CONTROL NUMBER (If FAA registration is not required for your aircraft, explain in item 15) N *1234*

8. FEE TYPE CODE	9. FEE MULTIPLE	10. FEE DUE	FOR FCC USE ONLY
P A A	*O O O I*	*$115.00*	

11. TYPE OF APPLICANT

- [✓] I–Individual
- [ ] D–Individual with Business Name
- [ ] P–Partnership
- [ ] C–Corporation
- [ ] A–Association
- [ ] G–Governmental Entity

12. PURPOSE OF APPLICATION

- [✓] New Station
- [ ] Renewal
- [ ] Modification (Specify) _____

13. IS APPLICATION FOR A FLEET LICENSE?
  A. If modifying a fleet license, give the number of aircraft to be added.
  B. If applying for a new or modified fleet license, give the total number of aircraft.

[ ] YES [ ] NO

14. FREQUENCIES REQUESTED (Check appropriate box(es) in 14A and/or 14B, see instructions)

A. CHECK ONLY ONE	B. ADDITIONAL INFORMATION IS REQUIRED IF YOU CHECK HERE	
[✓] A–Private Aircraft	[ ] T–Flight Test HF	[ ] P–Portable (Showing required)
[ ] C–Air Carrier	[ ] V–Flight Test VHF	[ ] O–Other (Specify)_____

15. ANSWER SPACE FOR ADDITIONAL INFORMATION

CERTIFICATION
1. Applicant waves all claims for the use of any specific frequency regardless of prior use by license of otherwise.
2. Applicant will have unlimited access to the radio equipment and will control access to exclude unauthorized persons.
3. Neither applicant nor any member thereof is a foreign government or representative thereof.
4. Applicant certifies that all statements made in this application and attachments are true, complete, correct and made in good faith.
5. Applicant certifies that the signature that appears on this application is that of a person with the proper authority to act on behalf of the party represented.

WILLFUL FALSE STATEMENTS MADE ON THIS FORM ARE PUNISHABLE BY FINE AND/OR IMPRISONMENT (U.S. CODE, TITLE 18, SECTION 1001), AND/OR REVOCATION OF ANY STATION LICENSE OR CONSTRUCTION PERMIT (U.S. CODE, TITLE 47, SECTION 312(A)(1)), AND/OR FORFEITURE (U.S. CODE, TITLE 47, SECTION 503).

→ **SIGNATURE** *Ima Loft*   **DATE** *3/23/95*

FAILURE TO SIGN THIS APPLICATION MAY RESULT IN DISMISSAL OF THE APPLICATION AND FORFEITURE OF ANY FEES PAID.

Fig. 8-2. Radio station license application.

Now here comes the interesting part. Sometimes the seller "neglects" to fill out the selling price section on the top of the bill of sale. The temptation to fill in a lower price is very tempting.

Fig. 8-3. Although it doesn't normally carry a radio, Fly Baby N500F has a radio license to allow occasional use of a hand-held radio to fly into controlled fields.

Fig. 8-4. Buying a new plane like this PZL Koliber II probably will require paying sales tax; however, many states have similar taxes covering private sales.

Be forewarned that the revenuers usually have aircraft price guidelines. A friend spent a nervous 15 minutes when the tax agent challenged the amount shown on a bill of sale. Fortunately, the aircraft in question was a homebuilt and wasn't listed in the tax agent's book. Fines and prison terms tend to interfere with your flying.

That's not to say you can't look for legal ways of reducing the bite. It pays to check around. Some states, such as California, don't levy a tax if the plane doesn't enter the state for at least 90 days after purchase. Your state aviation department might be a good place to ask for specific tax provisions.

Other paperwork

You can apply for a vanity registration number if you wish. Your options aren't quite as wide as those for cars. It must, of course, begin with N, and can't be more than five additional characters long. It must consist of at least one number and end in no more than two letters.

The drawback is that the old registration number is already permanently applied. For this reason, custom N numbers usually are the province of the homebuilders. Still, if you are restoring the aircraft (or just repainting) you can get a custom number. AC Form 8050-64 is used.

INSURANCE

From most aircraft owners' viewpoints, the insurance world is a vast jungle studded with pitfalls and sharp-toothed carnivores. There's some truth in that. Insurance is a rather esoteric field that most of us prefer to avoid.

Still, you're an airplane owner now. Grab a machete and be prepared to hack through some red tape. We talked a bit about insurance in chapter 2. Back then, I recommended that you call some agents and get some price estimates. Now that you have a real piece of hardware, call them back, and get definite quotes.

Before you do, let's examine the types of coverage, plus the loopholes and whatnot that lurk within insurance policies. One thing to keep in mind: Insurance policies vary from company to company. Subtle differences in wording can make a world of difference. Quiz the agent for anything that you don't understand.

Liability coverage

Generally speaking, the "standard" for liability *bodily injury* (BI) insurance is $100,000 per passenger in the aircraft, $100,000 per person on the ground, and $100,000 for property damage. Typically the insurer also places a cap on how much the company will pay if you are found liable for an accident; $500,000 to $1 million are typical amounts.

Are these levels right for you? With the way the courts seem to run, maybe you'd want higher limits. Take the $100,000 property limit, for instance. If you taxi into someone's new Bonanza, you might be on the hook for three times the covered amount.

There are a few ways to reduce the coverage, and hence your premium. If you just intend to fly family members, passenger liability coverage can be excluded or the coverage level reduced. Then again, it's not unknown for siblings or offspring to go to court, is it?

The liability premium will vary with how many seats the airplane has. Coverage of our club single-seat homebuilt costs only $215 a year; the owner of the seven-seat Cessna Stationair in Fig. 8-5 will pay far more. If your airplane can carry six or more passengers, you can sometimes reduce your premium by removing one set of seats, assuming that it's allowed by the aircraft's type certificate, of course.

Cessna Aircraft Company

Fig. 8-5. The insurance premiums for larger aircraft can sometimes be reduced by removing seats.

Some owners operate "bare," with no liability coverage. It's a tough course to take in this litigious age. Considering all the other expenses of ownership, the cost of liability insurance isn't that significant.

Just as a benchmark, I called a national insurance company and got a quote on myself. I'm a 500-hour private pilot without an IFR rating and no violations or accidents. For $1 million worth of coverage, the company quoted me $340 a year for liability on a Cessna 172.

Hull

Hull coverage is the equivalent of automotive collision or comprehensive coverage. Some aircraft owners don't carry it. They prefer to self-insure; if they break the airplane, they'll fix it as their funds permit. Deciding to self-insure is like betting with the insurance company. Fly 20 years without crashing and you win! It's not for the faint of heart or those with expensive airplanes.

There are several different varieties of hull insurance, depending upon what level of protection the owner desires. These levels can allow you to get the best coverage for the available money. From the least expensive on up:

- *Not in motion* policies cover the aircraft for physical damages that occur when the plane isn't moving and/or the engine isn't running. This is useful for protection against weather damage or vandalism. If hail smashes the windshield or dents the wings, the insurance pays.

- *Not in flight* policies cover the aircraft for physical damage up to the point where it taxis onto the runway for takeoff. The policy goes back into effect when the plane safely reaches the taxiway after landing. If you taxi too close to a signpost and ding a wingtip, the policy should apply.
- *In-flight* should cover your aircraft against risks on the ground and in the air. This coverage is sometimes called *all risk*, though that isn't quite accurate. All policies contain exclusions—cases where the insurance isn't valid.

The specific versions of these types of coverage vary between insurance carriers. It is *vitally important* to understand the "definitions" and "exclusions" listed in your own policy.

The coverage you select should depend on your level of comfort. Back when I owned a $6,000 Cessna, I carried "Not in Motion" coverage. I was willing to bet that my piloting skills were sufficient to prevent the need for an "In Motion" policy. However, if I paid $35,000 for a used Aviat Husky, I'd probably want full protection.

Back to the benchmark Cessna 172: I asked a major insurer for a hull insurance quote. The aircraft was a Cessna 172 to be tied down at a suburban uncontrolled airport. As I mentioned, the premium for liability coverage was $340. Adding "Not In Motion" hull coverage added another $200. The premium for "All Risk" increased the annual bite by $250. Hence, the yearly premium was about $790.

This issue of whether or not to carry "All Risk" coverage is moot if the airplane acts as loan collateral. The lending institution will require full hull insurance.

The perils of underinsurance

There's one way to economize on hull insurance coverage, but it can really bite. Let's assume you bought that used $35,000 Aviat Husky. If you insured it for, say, $20,000, the hull premiums would be less. Moderate damage, say a prop strike or a minor ground loop, would be covered just as well as if you carried coverage for the full $35,000 value of the plane.

But what if something *really bad* happens? Say a tornado hits and crumples one wing pretty severely. It doesn't take much to run up a high repair bill. The insurance company examines the plane and determines that the mangled Husky has suffered $16,000 worth of damage.

That $16,000 is 80 percent of the insured value. The company will probably declare that the aircraft is a total loss. They pay you the $20,000 that the policy is worth, and that's the end of that; $20,000 isn't enough to buy a replacement Husky. So you're out of an airplane, even though the amount of damage was less than half the actual value. Some salvage dealer will buy the wreck at auction for $10,000 or so, rebuild it for another $12,000 (they've got the parts), and sell it for $30,000, making an $8,000 profit that essentially came right out of your pocket.

Keep this in mind if you upgrade your airplane with new paint or an overhauled engine. These improvements increase the value of your machine and should be reflected in your hull coverage. Don't forget to have the insured amount track the actual replacement cost of the machine.

Overinsurance has its faults as well. Let's assume that for some reason you insured that Aviat for $50,000. The premiums would be higher, but at least if the plane is totaled you'll definitely be able to buy another one right away.

Then a windstorm hits the airport and blows the Husky across the field. The adjuster looks at the results and estimates $30,000 damage, which is almost what you paid for it. But the amount of damages is just 60 percent to the insurance company. The plane isn't totaled. They could hand that steel-and-fabric tumbleweed to the lowest bidder. You'll get an airplane back months later. It probably will never be the same, and you'll never be able to sell this MDH airplane for anything like you paid for it.

It won't necessarily happen that way. Some companies recognize when planes were overinsured and will negotiate a total loss figure. Still, it will cause fewer problems all around if you insure the plane for what it's worth. Don't forget to adjust the coverage as your plane's value changes.

Policy language

Now we come to the nitty-gritty. Here are some phrases or concepts that might be included as part of your insurance policy. Don't take these as universal because the same terms might be used differently on your policy.

Geographic exclusions
The policy might only be in effect when the aircraft is within a certain area. For instance, the coverage might apply only in "the 48 contiguous United States." If you wanted to fly to Canada, Mexico, or Alaska, you'd have to get an additional policy (*rider* or *endorsement*) to ensure coverage.

Named pilot(s)
Most policies are in effect only when specific pilots are flying. The remainder list the pilot qualifications necessary for the policy to be in force. If the pilot isn't named on the policy or doesn't meet the training/flight experience requirements, the policy won't be in effect. Most policies will cover pilots with commercial licenses test-flying your plane in conjunction with maintenance.

Pilot qualifications
The pilot must meet all the FAA requirements relative to flying the insured aircraft. If your medical or BFR isn't current, then your aircraft policy might be void. If the aircraft is "high-performance" and you don't have a high-performance endorsement, the insurer might refuse a claim.

The aircraft must be airworthy
If you groundloop and the annual is overdue, the insurance company might deny your coverage. If a "Not in Motion" policy is in force, this exclusion shouldn't apply. Ask your agent.

Better-quality policies don't include the last two provisions when the discrepancy cannot be shown to be *causal*, that is, contributing to the accident. In other words, if your medical certificate has expired, the insurance should be in effect unless some medical condition caused the accident or made the outcome more severe.

Compliance with Federal Aviation Regulations

This provision is loaded with dynamite; don't accept it on your policy if it can be avoided. Most accidents are accompanied by a FAR regulation. FAR 91.13, "Careless or Reckless Operation" can probably be applied to just about any instance when aluminum is bent.

Legal or illegal seizure

If you use your Cessna 180 to haul happy dust into the country, the insurance company doesn't compensate you if the Feds seize your airplane.

Betterment

To an insurer, *betterment* is when the repairs after an accident leave the aircraft in better shape than before. That's a no-no. Say your brakes lock on landing, blowing two tires and causing some wheel and brake damage. Your insurer will look carefully at those tires. If they'd been worn halfway prior to the accident, the insurance company might only pay for one-half the value of the new tires.

If your constant-speed prop hit something while the plane screeched to a halt, the same applies. Some models have to be torn down and overhauled every five years; if your prop's time was up, the insurer might refuse to pay for it.

Engines are a bit different. Good hull coverage pays for post-accident teardowns and inspections in *sudden-stoppage* cases, generally regardless of engine time. And the insurance will pay for damage caused by the accident, but if the inspector finds a spalled camshaft, replacement comes out of your pocket, not the insurer's.

Here's another example: You're flying along and the engine quits. You do a semisuccessful forced landing. If the engine quit because the crankshaft broke, the insurance *will not* cover its repairs. That would be betterment. If the failure was due to a clogged gas vent, and the engine was damaged on impact, the insurance *will* cover the engine repairs.

Breach of warranty

If you took out an aircraft loan, the lender required hull coverage to protect the loan's collateral. If you look on your insurance policy, you might see the term "Breach of Warranty."

A breach of warranty clause requires the insurer to pay the coverage amount to the lender *even if the policy is otherwise voided*. So if your BFR is overdue and you overshoot the landing, the insurance company still pays off the loan. Home free? Not hardly. The insurance company will sue you for the amount that they had to pay your lender, a process called *subrogation*.

Again, check over the policy thoroughly to be sure the provisions are acceptable.

The criticality of compliance

It is important to *exactly* conform to the insurance company's requirements. Insurance companies make money by collecting premium payments, not paying off claims. If you buy a retractable-gear policy and the insurer requires "Owner must log 15 hours of dual before hull coverage commences for solo flight," make *darn* sure that at least the required amount of time is *entered* into your logbook prior to flying solo.

If you take your 15th hour of dual and forgot to bring your logbook for the instructor to enter the lesson, *don't go for a celebratory solo jaunt around the pattern*. Remember, the requirement is for *logged* time. Otherwise, you'll just be giving the insurer grounds for denying any claim.

The same situation holds true when applying for coverage. The insurer might not scrutinize your application when your coverage starts, but might examine it closely if you file a claim. If you claimed 1,000 flight hours, but your logs only show 500, that's grounds for denial.

If you're carrying "Not in Flight" or "Not in Motion" hull coverage, understand *exactly* the insurer's definition of these terms. "Not in Motion" coverage should have some nice legalese along the lines of "Motion imparted by engine thrust or as a result of flight."

My friend Jim once had a passerby "help" push his plane into the hangar. The overmuscled stranger shoved it right into one of the roof supports. What if the hull coverage had only said "Not in Motion"? After all, the plane *was* in motion at the time of the accident.

If you don't understand any term on your policy, get the agent to explain it to you. Ignorance is no excuse for noncompliance.

Starting the coverage

If you work through a local agent, insurance should start upon your delivery of a check for the premium. Most companies have payment plans, so the entire amount might not be immediately due. Some companies have monthly plans, some quarterly, and others semiannually. Again, it's something to ask about before the actual aircraft purchase.

To minimize the premiums, become a better pilot. Get advanced certificates and ratings, especially an instrument rating. Participate in the FAA's Wings program. If you're carrying hull coverage, put the plane in an enclosed hangar. Remember, you'll be paying quite a bit for your insurance coverage. Make sure you don't negate it by violating some term within the policy.

TYING DOWN

The plane's yours, but you can't just let it sit around loose. You'll want to tie it down before you go home that first day. Maybe you've got a closed hangar, and figure you can skip this part. Look at your lease: It might require you to tie the airplane down even inside the hangar. That's the way it is at my home field.

Most of us are going to have to secure the airplane. There are two elements to this: the anchors and the tiedowns.

Anchors

If the tiedown spot is paved, the anchors will consist of an embedded steel loop, hook, or eyebolt. There's usually not much you can do to improve these anchors. Take a look, though; make sure they're solid.

Grass parking areas have a variety of anchor systems. Some have a long cable strung across the tiedown area, tacked down at intervals. Or each spot might have an individual anchor.

Cast a jaundiced eye at the anchors on turf. Talk to the airport manager; the tiedowns are usually the complete responsibility of the owner. Those at your spot might have been left by the previous occupant.

You then have your choice of using it as-is or installing your own. A pull test might be a start, but you'd like for each anchor to withstand a thousand-pound pull, which is tough to verify by just tugging on it.

Otherwise, you can install your own anchors. You need something that goes in deep and grabs onto a wide patch of dirt. Aircraft supply catalogs sell tiedown kits, but you might do better at a local farm-supply store. You could even dig a deep posthole, insert a hooked steel rod, and fill it with cement.

You'll need two anchors set apart at the approximate distance between the wing's tiedowns, plus one a few feet in back of the tail. Some people use concrete blocks to tie down the tail, but I'm pretty skeptical about that arrangement.

Tiedowns

With the anchors established, you'll need something to connect them with the aircraft's tiedown points. You'll generally see one of three systems used: rope, chains, and straps.

Rope tiedowns are probably the most common. They're inexpensive to buy, easy to attach, gentle on the plane, and easy to tighten. They can have a problem with long-term exposure to the environment. I once tried to tie my plane down at a transient spot at a remote field. I grabbed onto one of the ropes, leaned backward, and it broke. So did the other two. Ultraviolet light had done them in.

You may choose between three general materials: manila, nylon, and polypropylene (commonly called "poly"). Manila will be the cheapest. It's rather poor at strength and abrasion resistance, and moisture causes deterioration. It'll stretch and shrink with the weather. You'll have to learn how to do a little bit of rope splicing to keep the ends from unraveling. It withstands sunlight a lot better than synthetic ropes.

Nylon and poly are the most common. Poly gets a lot of attention in the boat world because it floats, but that's neither here nor there for aircraft. Poly is a bit more susceptible to abrasion damage and deterioration from sunlight than nylon.

You'll want about ½-inch-thick rope. For nylon, that corresponds to about 4,000 pounds capacity. Its "working rating" is usually 10–20 percent of its capacity. A half-inch nylon rope is good for about 600 pounds.

The ends of nylon and poly rope will tend to unravel. After they're cut, get a match or other flame source. Move well away from any aircraft and light the end of the rope. Let it hang down and burn for 10 seconds or so, then blow it out. Let it cool. The end of the rope will have melted together. A car's cigarette lighter can be used instead of a flame, but it might take several applications. The lighter stinks a bit afterward, too.

You won't have to do the same trick with a chain, but you will need a hacksaw instead of a penknife to cut it to length. You'll also need D-links or something to attach the bottom of the chain to the anchor, and a clip for securing the upper end.

Chains are, at first thought, a natural for aircraft tiedowns. They're strong and don't deteriorate; however, they're tough to tighten. If the links are each an inch long, you can end up with that much slack in each chain. If a wind rocks the plane, the chain jerks it to a stop. That jerk can induce serious loads; in extreme case, you could damage the tiedown point. Rope can be adjusted to exactly the right length. And it has a little "give," which reduces the strain in gusty winds. A chain can also scratch the aircraft's finish. Always secure the free end so that it won't swing in the wind.

Finally, there are web or strap tiedown systems. By themselves, they are similar to nylon or poly rope. Some companies sell systems with tensioners that make it easy to take the slack out. I don't really have any experience with these, but they probably work OK. It's just more expensive than rope.

How to get tiedowns tight

I can't tie a knot worth a darn. There are some ways around this. I use a little homemade tightener; similar units are available commercially. But there is a way to ensure snug ropes without any additional hardware. All it needs is the tail anchor to be a bit behind the aircraft.

Park the airplane with the wing tiedown points directly over the anchors. Make sure the parking brake is off. Tie the wings down as tight as you can. Go to the tail and put the rope through the eyelet, or around whatever other tiedown point is provided. Then pull on the tail rope to tug the aircraft backward. The wing ropes will tighten up. When the plane stops moving, tie off the tail rope. This process is summarized in Fig. 8-6.

AIRPORT ETIQUETTE

Up until now, you've had a pretty easy flying life. The FBO probably had the plane ready on the ramp when you came out to fly. They probably handled the fueling and preparation. All you had to do was preflight, hop in, start, fly, and park it in the spot you left from.

As such, you probably haven't had much of a chance to inconvenience or anger someone else. And even if you did, you were flying an anonymous rental aircraft, right?

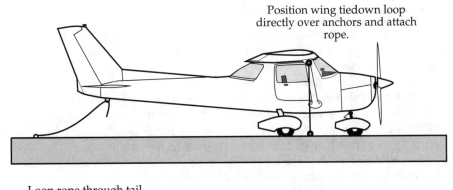

Position wing tiedown loop directly over anchors and attach rope.

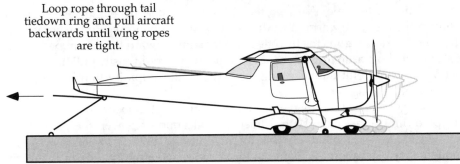

Loop rope through tail tiedown ring and pull aircraft backwards until wing ropes are tight.

Tie off tail rope, chock wheels.

Fig. 8-6. Tight tiedowns are inexpensive insurance.

That's changed now. You're going to be "living" next to a bunch of other aircraft owners. They're a swell bunch of people, mostly. They're willing to bend over backward to help a fellow owner. You'll be able to pick up a lot of advice on maintenance items, insurance, mechanics to avoid, and a lot of other subjects.

These owners will only ask one thing: that you don't do anything that might damage their own aircraft or interfere with them unnecessarily. To prevent you from unnecessary faux pax, let's discuss some airport etiquette.

Watch your prop blast!

I talked to a lot of pilots about what really got them mad. Number one on the list, by far, is owners who don't minimize the propeller wash hitting other aircraft. It seriously ticks people off. Believe me.

Even at idle, propellers shove a lot of air backwards. Prop blast flings gravel on paint jobs. It shakes planes off jacks. It blows cowlings across ramps. It knocks over ladders. It catches aircraft doors and slams them against the stops.

There are some incredibly stupid people out there. Two hangar rows from me, there's a guy with a Cessna 182. When he goes to fly, he rolls the plane out of the hangar and turns it 45° to the right. The tail is pointing *directly* at the award-winning fiberglass homebuilt airplane in the next hangar.

And time after time, I've seen the 182's owner start his engine and sit there idling until his engine warms up. All the while, gales of wind scour the homebuilt. Then the 182 driver guns the throttle to start taxiing toward the runup pad. There is absolutely no reason for such behavior. The owner could easily pull the plane 15 feet farther and point the tail along the taxiway. It's stupidity in this case, "hot-dogging" in others.

One advantage of taildraggers is that the tailwheel unlocks, and the airplane can pivot in a tight circle. With a liberal application of power, the plane fairly pirouettes. I've seen pilots do this to turn around on narrow taxiways. Sure enough, they blast about six planes in the process, and they seem to think that the glances shot their way are in admiration.

You should have the idea by now. Here are some specific rules to live by:

- Never, *ever* start your engine with the tail pointed at someone else's airplane. Roll the plane out of its parking spot and turn the tail to point down the lane between aircraft. If it's impossible to avoid—you're parked in a U-shaped formation of airplanes at a fly-in—first roll the plane forward by hand as far as possible. Ask bystanders for help; most owners will be glad to assist.

- Use no more power than is necessary. Even with the tail pointed down the taxiway, propwash spreads.

- Use momentum when you can't avoid pointing the tail at someone. For instance, if you have to turn to the side to point the tail at your hangar, let speed build a bit and pull the mixture early. Make the final turn as the engine winds down. It might look like hot-dogging, but surrounding owners will bless you for it.

- Finally, not every pilot likes aircraft noise. If you need to run the engine for a while to warm it up for an oil change, taxi to an isolated part of the field. Don't just roll it onto the taxiway in front of the hangar and run it up. (Consider some touch-and-goes to simultaneously warm up the oil, check out the airplane, and ensure currency.)

Don't block the hangar or taxiway

Many airports have a rather narrow lane between tiedowns or hangars (Fig. 8-7). Yet there are always owners who narrow the space farther by parking their cars in front of their hangar, or roll their plane partially onto the taxiway then drive away for lunch.

Fig. 8-7. Most pilots are nervous when taxiing in tight quarters. Don't make it worse by parking your car in the wrong spot.

And even if you tuck your car off to the side, make sure that you don't block anyone's hangar door from opening all the way. The person in the hangar next to me used to leave his car parked along the wall between our two doors. He'd position it so that his hangar wouldn't be blocked, yet the front of his car would encroach the door of mine.

One time, another man came around the corner to find me and a couple of other guys *lifting* a wing over the hood of his pickup truck so the plane owner could put his machine away. The truck owner was apologetic, but that didn't make him any more popular.

When we drive cars, we have a pretty good feel about the location of the fenders and the space occupied by the car. A pilot who is taxiing an airplane doesn't like to take risks. When you park your car at the hangar, it might look like any plane has plenty of room, but things look tighter from the cockpit.

Keep these points in mind:

- Never park your car where there's any possibility of impeding aircraft traffic.
- Never park your car directly across from another parked car or other obstruction.

- If you're going to be gone for a while, always park in your airplane's spot (assuming the airport allows this practice). Parking inside the airplane's hangar or covered spot is all the better.

- Never immobilize your plane (take the wheels off, etc.) in a position where it could interfere with traffic.

- Don't push the plane onto the taxiway until ready to start the engine. Do your preflight at the tiedown or in the hangar.

Ask before touching

In the wild west, you didn't climb down off your horse on someone's ranch until you were invited. Similarly, don't enter someone's hangar without asking first. It doesn't take anything elaborate: "Hi! Nice plane. Mind if I take a look?" Most owners won't mind, but they'll be a little less hostile if they feel you're careful around their aircraft.

Sure, if it's a big open hangar with many airplanes right next to each other, you don't really have to ask just to walk up to an airplane. For courtesy's sake, you should still ask permission before moving in for a close examination.

It might pay off. Our club's little open-cockpit homebuilt sits in an open hangar, and a lot of people wander by to look at it. I really have no problem with that. A number of times I've looked up from my work to see someone leaning over the wing to look into the cockpit. Again, no problem, but sometimes someone will first ask "Mind if I take a look at it?"

Those that do ask are invited to sit in the cockpit. Because they are cautious enough to ask, I figure that they probably can be trusted clambering over the cockpit sides.

On a related note, though, one should *never touch an airplane without permission*. Don't wiggle the control surfaces because the owner might be working on them. Don't "plonk" the fabric because it causes "ringworm" ridges. Don't stick your nose into where a lot of little goodies, especially tools, are lying loose; the owner might think that you are trying to steal something.

Above all, don't "help" move an airplane without being specifically invited. Owners prefer to keep full control, especially in tight situations. Plus, you might pick the wrong place to push.

Other peeves

Keep kids restrained and pets on leashes (Fig. 8-8). A dog that runs out to chase an airplane might accidentally run into the prop. And the antennas and pitot tubes jutting out from every airplane are often well-placed to get bent or broken by a running five-year-old. One man found a child riding piggyback on the top of his Champ's fuselage with the proud parent snapping a photo.

Turn the strobes off when around hangars or tiedown areas at night. They can blind someone trying to secure or preflight their aircraft.

Fig. 8-8. When taking your pet human for a romp on the airport, bring pretzels and a leash.

When you drain fuel during preflight, don't dump it on asphalt. Asphalt is petroleum-based; gasoline eats it away. Concrete is better. Explore replacing clean, uncontaminated fuel to minimize or eliminate pollution at the airport. Never light up a cigarette in the vicinity of any airplane; watch friends, visitors, and strangers very carefully to prevent unintentional carelessness with an open flame.

Yell "Clear!" so people in nearby hangars can hear you, and wait a moment to let them secure any loose items. Never leave a stepladder upright because it could get blown over and tumble into a parked airplane.

Uncontrolled airport guide

Perhaps you're like many pilots, and you learned to fly from a field with a tower. You undoubtedly make some flights to nearby uncontrolled fields, but most of your operations have some level of ATC interaction.

It's quite possible, though, that you might end up based from an uncontrolled field. Tiedowns and hangars are generally cheaper, and an uncontrolled airport is often convenient to the suburbs where most of us live. There are a few points to remember around uncontrolled fields:

- Runway space is gold, especially on sunny summer weekends. After landing, leave the runway at the earliest safe turnoff; safely add power to reach the next turnoff, if necessary.

- Don't block access from the taxiway to the runway during your runup. Some aircraft have very short pretakeoff checks. Most fields have a runup area; use it.

- Many uncontrolled airports have problems with the neighbors. Don't make matters worse. Do follow noise-abatement procedures.

- Don't "abandon" your plane at the gas pumps. After filling up, get clear so other planes can fuel.

- About seven percent of the aircraft in the United States don't have radios. Few of these are based at tower airports, so the proportion is higher at uncontrolled fields. Be on the lookout. Remember that they cannot hear your announcements on the unicom/CTAF, so avoid nonstandard procedures such as 360° turns on final.

- Straight-in approaches aren't prohibited by regulation, but many pilots are strongly opposed to them. You'll make no friends if you make straight-ins a habit.

We survived the first week of ownership. It's time to deal with the long-term.

9

The first year and beyond

YOUR NEW PLANE IS GOING TO REQUIRE some involvement on your part, beyond just operation. Whether or not you plan to involve yourself in its maintenance, there are a number of other details that you'll have to attend to. This chapter covers issues that will or might arise in your first year of ownership.

You might not actually face some of them within your first 12 months, but they might come up at some point. We'll cover them here for convenience. One thing that will arise in your first year is the annual inspection, and that is specifically covered in the next chapter.

FEATHERING THE NEST

When you've got your plane settled into its new nest, you'll want to make that hangar or tiedown spot as "homey" as possible. Let's examine some of the things that'll make your flying life easier.

Chock talk

"Chocks?" you say? "Why use chocks when the plane has a parking brake? Besides, the tiedowns will keep the plane from rolling." Why not use the parking brake? Here's a good reason: Say you're in a row of hangars with a number of other aircraft and one catches fire. The fire department decides to move your plane away from the fire zone. The action has nothing to do with any deep consideration of your investment; they just want to keep the loaded fuel tanks from adding to the conflagration.

Firefighters undo the tiedown ropes, simple enough. Nothing blocking the wheels. Then they pull on the plane. It doesn't roll. Do you think they're going to try to figure out why? Heck no! They'll throw a log chain around the prop or a strut and drag the plane away.

Chocks are handy things. If your parking spot slopes a bit, they'll stop the plane from rolling while it's being tied down. They'll keep it from moving when you jack one side up to change a tire. And if the plane doesn't have a starter, they're just about required during handpropping.

Chocks are another item that you can spend a little or a lot of money on. There are nice lightweight travel units available through the aircraft catalogs. If you've got a table saw, you can make some units with a nice bevel, but even a cheesy handsaw can make some nice serviceable chocks from three bucks worth of lumber (Fig. 9-1).

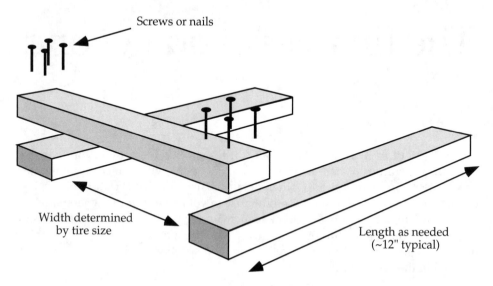

Fig. 9-1. You can make a simple set of chocks out of a couple pieces of lumber.

Control locks

Hopefully your plane came with control locks. Letting the wind knock your ailerons and elevators around wears out the system's bearings and bushings. Closed hangars solve the problem, too, but control locks are cheaper.

You usually can secure the stick or yoke with one of the seat belts. There are drawbacks, though. At least one control is usually held hard-over. It'll add to rocking; if the wind gets really strong, it could even contribute to blowing your plane over.

At a minimum, pick up a set of the manufacturer's cabin locks. These, or similar units, are available through many suppliers.

The best solution is a set of external locks on the control surfaces. These have the added advantage of keeping any strain out of the control system. Yoke or stick locks still allow the ailerons or elevator to strain against their hinges and cables. External locks can be expensively purchased or cheaply made from lumber, using old carpeting for padding (Fig. 9-2).

Towbars

If you've got a tricycle-gear aircraft, you should get a towbar, which is relatively inexpensive at $20–$40. You might think that you can get by without one.

Fig. 9-2. Exterior control locks are the best for preventing damage.

Some owners maneuver their Cessnas around by pushing down on the horizontal stabilizer to lift the nosewheel off the ground, which is a bad idea.

Some older 150s and 172s have developed cracks in the stabilizer's forward spar, and ground handling has been getting some of the blame. It's not something that occurs suddenly; nothing will probably happen the first time that you shove your 172 into the hangar. Long-term abuse of this sort seems to crack the spars around the bolt holes.

While a replacement spar is a seeming bargain, less than $300 at this writing, it takes quite a number of labor hours to install. The total bill for one 150 owner came to over $1,200. That $25 towbar is a little easier to justify than $1,200.

Other aircraft accessories

How's the bird problem in your area? The feathered birds, not the metal ones. (Odd, now that I think about it. In 20 years of flying, I've never had a bird nest in a cowling. Probably just lucky.)

Closing off the openings in the cowl helps prevent bird problems (Fig. 9-3). A number of companies make custom plugs for aircraft cowlings. There are outfits

Fig. 9-3. Cowl plugs help keep birds out of the engine.

who sell do-it-yourself kits, too. Or you can buy a quarter-sheet of plywood or a hunk of 3" or 4" foam rubber and start hacking away.

One note of caution: Foam rubber does tend to disintegrate when exposed to weather. Picking little bits of foam from between engine cylinders is poor entertainment, so watch your plugs' condition.

Mud dauber wasps are good reasons for pitot and fuel vent-tube covers. Again, these can be purchased ready-to-use, or you can buy a little vinyl and try to figure out the family sewing machine.

Commercial cowl plugs and pitot covers usually come with long red "Remove Before Flight" banners. Add something similar to homemade units. Your checklist should also have a similar reminder in print: "Remove all gust locks, plugs, and covers prior to engine start."

You probably won't forget to remove windshield or instrument panel covers. These range from $5 "space blankets" thrown over the instrument panel to custom-fitted exterior covers (Fig. 9-4). They have two major uses. First, in hotter climes, they can reduce the interior temperature. It's not just a comfort issue; avionics and upholstery will last longer. Second, a windshield or panel cover hides your expensive radios from those with a larcenous bent. Thieves won't usually break into a plane unless they know exactly what prize awaits.

Custom-fitted exterior covers are probably the all-around best. In addition to protecting the interior, they also keep the Plexiglas dirt- and dust-free between flights. But because they're on the outside, radio thieves can easily peek underneath to see what the panel's packing. External covers also take a beating from the

Fig. 9-4. A custom-fit cockpit cover is the best choice; however, it will be the most expensive option.

weather and must be replaced occasionally. Then again, it's better to replace your cover than your windshield.

Speaking of beating, pay close attention to *how* the covers attach. You don't want them damaging your paint. All grommets should be isolated from the exterior; all straps should have smooth exteriors. The straps and grommets should hold the cover securely and keep it from flapping. A flapping cover can bang on the aircraft skin and rub away paint, which is one drawback to the $1.99 blue plastic tarps that are thrown over some planes; they can't really be cinched up tight. Still, these covers might be your only option if rainwater keeps leaking into the cabin. Just be careful how you arrange the bungee cords (Fig. 9-5).

A second sort of cover has a shiny surface and attaches to the inside of your windows. While this cover doesn't help protect the Plexiglas, it does keep the temperature down and keep the panel hidden.

Hangar goodies

Accessories and gadgets can make airplane ownership easier. Chapters 11 and 12 discuss owner maintenance and required tools. Here's a few items that you might find useful now:

- Engine oil. You can buy oil a lot cheaper by the case from an aviation discounter than quart-by-quart at the FBO. In any case, if the engine oil is low, access to your own oil avoids a side trip.
- Extra inspection panel screws. Take one of the screws that holds inspection

Fig. 9-5. Impromptu covers might be better than nothing. Make sure that they are solidly secured to prevent damage to the aircraft paint.

panels to your local aviation supplier. Buy 10 spares and put them in an old film container. If there isn't a nearby aviation hardware source, order some by mail.

- A screwdriver assortment. You'll need them for the screws. A cheap set of screwdrivers (suitable for occasional use) can be bought for fewer than $10.
- Plexiglas cleaner and polisher. Airspace is crowded; keep the windows clear.
- Air pressure gauge. The regular automotive kind is fine, although if you've got a Cub or other plane with low-pressure tires, make sure the gauge will give readings below 10 psi.
- Tire pump. Aircraft tires don't carry much air, compared to car tires. A simple bicycle pump is sufficient for adding a bit of air to a low tire. If you're lazy like me, pick up one of those 12-volt car emergency tire pumps; if your airplane does not have a 12-volt system, you will have to run the pump off your car.
- General-purpose cleaner. A spritz of household cleaner on a fresh grease spot keeps the stain from becoming permanent.
- Waterless hand cleaner. Sometimes you'll get a little dirty working around the plane, and a container of Go-Jo or equivalent cleaner keeps you from leaving grease marks on your car's steering wheel.
- A jug of water. The waterless cleaner does leave your mitts feeling a bit funny; sloshing water on them really helps.
- A small glass jar or two that you can shoot fuel into to take a look at it, or use to test autogas for alcohol (explained later in this chapter). Food jars are fine.
- Paper towels and/or shop rags for wiping and cleaning.
- Old newspapers. Use them to catch spilled oil, and the like.

It's ideal if you can arrange a way to store this stuff out at the airport. Check on your lease and see if it's OK to place a cabinet in your hangar. Don't run off and buy a brand-new locker. Watch the garage sales and auctions. I picked up a 6-foot-tall cabinet for $12 at a local surplus store. Old school lockers work just fine, though perhaps are a bit narrow.

You probably won't be able to put one at an open tiedown. The airport management might allow small footlockerlike units. I've seen people use those large plastic containers designed for pickup trucks. A chain and padlock secures them to a tiedown anchor. The same hardware can secure a stepladder as well.

Finally, if you have a hangar to keep it in, a bit of carpeting or a large piece of cardboard make working under the airplane more comfortable.

Portable tiedowns

You're going to be flying to all sorts of fascinating places. Some might not have tiedowns available; it'll be a good idea to put together a portable tiedown kit. You'll need three nonpermanent anchors and some rope. *Don't even think about using straight spikes that resemble tent pegs.*

The usual anchors are large corkscrews, the longer and wider the better. Screw them into the ground at a 45° angle. They should lean away from the aircraft's tiedown point, as shown in Fig. 9-6.

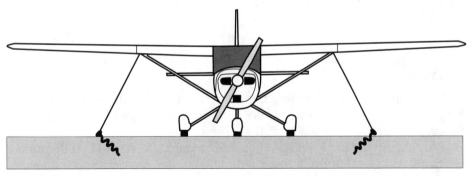

Fig. 9-6. Portable tiedowns should be inserted at an angle as shown.

You can buy such a tiedown kit through the aviation vendors, but the same corkscrews are sold in hardware and pet stores at nonaviation prices. I picked up three for less than $10. For storage and carriage, I screw the anchors into a piece of scrap styrofoam and keep the whole shebang in an old GI knapsack (Fig. 9-7).

KEEPING IT CLEAN

Pride of ownership will make you want to keep your airplane clean. Getting the goo and gunk off the paint will help it last longer, especially if you wax it at the same time. Plus, a wash-and-wax job is a perfect excuse for a detailed exterior examination.

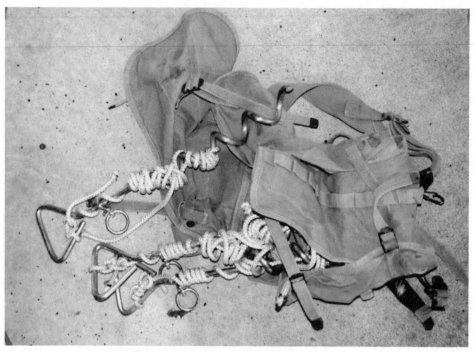

Fig. 9-7. The author's portable tiedown kit contains three "pet anchors" and some nylon rope stored in an army-surplus knapsack.

Washing

Most airports have a wash rack that often is just a spot with a faucet and hose. Bring a trigger-type hose nozzle. Many racks have them, but they seem to disappear. Bring your own and take it back with you.

Washing an airplane is similar to washing a car. Spray it to loosen the surface dirt. Scrub with a soft brush or towel with soapy water, then rinse off thoroughly (Fig. 9-8). Work in sections, rather than lathering the whole plane at once. A sample sequence is shown in Fig. 9-9.

Start with the right wing, washing it in 3-foot sections. Then divide the fuselage into four sections and do each individually: from the nose back to the leading edge, the cockpit section, the fuselage to the rudder, then the tail and rudder itself. Last comes the right-side horizontal stabilizer and elevator. On the fuselage and tail, work from high to low to chase the dirt downward rather than over just-washed areas. The process repeats on the left side of the plane.

You'll need soap or detergent in a bucket of water. Here's where a few problems might occur. Some paints, such as the "wet-look" polyurethanes, can be dulled by the wrong kind of cleaner. Most aviation supply houses will stock cleaners that are designed for all sorts of aircraft paint.

Fig. 9-8. Use plenty of water to avoid scratching the airplane's paint.

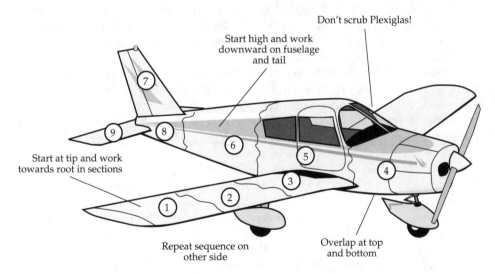

Don't scrub Plexiglas!

Start high and work
downward on fuselage
and tail

Start at tip and work
towards root in sections

Repeat sequence on
other side

Overlap at top
and bottom

Fig. 9-9. This typical wash sequence is an excellent opportunity for you to closely examine the airframe.

For those whose aircraft are painted with more traditional finishes like enamel, suitable detergents are no farther away than the local auto parts store. I use a "grease-release" dishwashing liquid on our club's fabric-covered airplane.

As far as applying the soapy water, just about any sort of soft brush or towel will do; either one must be clean, of course. A long-handled brush is mandatory to extend your reach to every part of the airplane, especially on a high-winger. It would be useful on low-wingers, too, by cutting down on the leaning and stooping.

To wash a section, first blast it with the hose to knock away any loose dirt. Be careful around the pitot/static system and any engine/spinner openings, though. Get the brush sopping wet in your water/soap mixture, and scrub the area clean. Resoak the brush often to keep the area wet because a dry brush drags dirt across the paint, adding tiny scratches that dull the finish.

Speaking of scratches, leave the Plexiglas alone at this point. Hose it down, but don't scrub on it. We'll do it separately later.

How big a section should you wash at a time? Well, you don't want any dirty, soapy water to start drying on the paint. It's actually more of a problem on the fuselage because most of the water runs away quickly. You want to rinse it clean with a blast from the hose before the dirty film has a chance to dry.

It'll take a bit of technique around the top and bottom. On the top, there's always a possibility some of the wash water will roll down the other side, which might have already been washed. Check often, and give any streaks a blast of the hose.

Plexiglas

The primary secret of Plexiglas cleaning is water, lots of it. You want to wash away any fine grit before it has a chance to scratch the plastic. When you wash the rest of the plane, blast the windows with the hose, but don't touch them with the scrub brush.

When you're done with the main washing, dump out your bucket, flush it with the hose, and fill it with fresh water. Get a nice soft *clean* sponge. Soak the sponge, then gently wipe it down the Plexiglas. Don't scrub it in a circle; don't rub it back and forth. Just start at the top and slide it gently down to the bottom. Shift to the side just enough so there's a little overlap with the previous swath and repeat.

When done, hit the windows with a Plexiglas cleaner. Most people use the stuff available through any aviation supplier. Stick your nose close to some windshields, though, and you'll catch a scent of citrus; some owners swear by "Lemon Pledge." Your call.

The main point is to make sure everything is clean. Set aside a sponge just for cleaning the glass. Mind your hands, too. I've never been able to wash an airplane without getting three-quarters soaked. Wet skin or clothing tends to pick up any dirt or sand it comes in contact with, so clean yourself off before doing the windows.

Some people say the interaction between paper towels and Plexiglas causes a buildup of static electricity. This charge then tends to attract dust. Rubbing a paper towel on the windshield *does* seem to build a charge. Pilot outlets sell static-free wipers for about $5 for 150 sheets.

The belly

I have a terrible confession. I generally wash the belly with the rest of the airplane, but make no special effort to scrub it clean. It haunts me at annual time. I end up flat on my back, cursing and scrubbing to get a year's accumulation of oily dirt off the belly.

There are a couple of alleviating factors in my case. First, the exhaust pipe of our club Fly Baby is long enough that the exhaust doesn't stain the fuselage. Second, the crankcase breather tube (our primary source of belly oil) is positioned to reduce the problem. With a production airplane, you're pretty much stuck with the way Mr. Cessna or Mr. Piper sold it.

You can buy aircraft cleaners specifically designed for cleaning exhaust stains and such from the belly. Again, find one compatible with the aircraft's paint. *Be sure to follow the dilution instructions because some of these products must be diluted 20-to-1 before use!* Sure, it'll clean better at the higher concentration, but stripping the airplane down to bare aluminum is probably not your first choice.

The instructions generally call for these products to be sprayed on and allowed to sit for a while. Garden shops sell pump-type sprayers, or you might be satisfied with the rather anemic output of the grocery-store spray bottle.

A more common option is to eliminate the middleman and use household cleaners in the spray bottles they're sold in. I've had especially good luck with a product called Simple Green. Previously, using common cleaners to get the oily dirt off the belly required a remarkable amount of elbow grease. Simple Green allows single-wipe cleanup. My local warehouse store sells refills in gallon jugs; my last purchase has lasted three years. Unfortunately, it doesn't seem to be marketed everywhere.

Whichever type you use, *read the label first* for any cautions regarding use on aluminum.

Other

Landing gear parts, such as oleos, tend to have a little seepage of hydraulic fluid that makes them dirt magnets. The grit chews on O-rings and seals, making the problem worse. Clean off the greasy parts with your cleaner. If it isn't quite strong enough, try petroleum solvent, mineral spirits, or a grease cutter called Varsol. Gasoline works too, but it's not a safe practice. If you must, wad up a paper towel, hold it under one of the airplane's fuel drains, and put a quick shot of gas on the towel.

The interior of an airplane can be cleaned just like a car. Aircraft carpeting is generally designed to be easily removable (it usually has to come out at annual time), so it's easy to vacuum up the dirt. Be very careful when using cleaners in the cockpit because windows are Plexiglas on the inside, too. Cleaner overspray can too easily drift onto a window and might cause permanent damage.

Waxing

Waxing the paint makes it look better and last longer. I've used automotive wax with acceptable results, though there are some deluxe silicone aircraft polishes that aren't really that much more expensive. The car-type waxes that I've used tend to get caught in seams when wiped on; then I have to dig it out.

If you've got a fabric-covered bird, be careful before you try traditional wax. The residue can get lodged in the fabric weave and be the very devil to get rid of. It isn't a safety issue (the weave just happens to show through the paint), but it looks horrible. Try it on a small, out-of-the-way area first.

Engine cleaning

Cleaning the engine isn't as simple as it once was. It used to be that you could take off the cowling, wrap plastic bags around the electrical stuff (magnetos, etc.) and vacuum pump, and blast away with engine degunker.

Environmental regulations are starting to limit the practice. My home field won't let us clean the engines at the wash rack anymore, and you really would like to flush the engine down with water after the degreaser does its thing.

If your airport isn't as restrictive, you can use the engine cleaners and solvents sold at auto-parts stores. A pump-type sprayer puts it on; let it soak for a bit, then flush it with clean water.

Due to my airport's limitations, I've taken to using Simple Green and paper towels. It's slow and tough to get into the tight corners, but it works. The process also gives me a slow, detailed look at the condition of the engine.

Cleaning the engine is also required at annual time. If you fly only 50 or so hours every year, your engine probably won't get dirty enough between annuals to require more than just a bit of wiping down.

HANGAR SHARING

If you own a simple aircraft like a Cherokee or 172, hangar rent can represent two-thirds of your fixed costs, assuming you can find a hangar. Sure, an outside tiedown is all right for an all-metal bird, but what about ragwing airplanes or that water leak you just haven't been able to eliminate?

Real birds share nests, why not aluminum "birds?" Squeeze two planes into a single hangar (Fig. 9-10) and you can slice your fixed costs by a major percentage. Your pride and joy can stay warm and dry for not much more than the cost of an open tiedown.

It's not as unwieldy as you might think. General aviation hangars are over-sized in order to fit a wide variety of planes. A pair of two-seaters, or a 172-sized plane with a homebuilt, probably can be made to fit. It might amount to a three-dimensional jigsaw puzzle, but careful planning and trial-and-error will yield a solution in a surprising number of cases. It depends on your willingness to put up with a little hassle in order to save money.

The legalities

Before you get your hopes up, check which local airports allow sharing. Some leases and city ordinances allow only a single aircraft per hangar. Some allow sharing, but require the second airplane to pay the equivalent of a monthly tiedown bill.

Fig. 9-10. Owners of this Cessna and Piper have reduced their hangar cost by 50 percent.

Finding a prospect

If you're OK legally, a prospective partner might be found via airport bulletin boards. If you already have a hangar, check with the names near the bottom of the airport's waiting list.

Try to select a partner with an airplane that maximizes your chance for success. Overlapping configurations and good ground maneuverability are key ingredients. Opposite wing arrangements (i.e., one high-wing and one low-wing) are best. Conventional gear aircraft are easier to maneuver than trigears.

Check with your local EAA chapter, too. Homebuilts are typically smaller than production aircraft (Fig. 9-11), and their owners are more used to handling aircraft in tight quarters like basements and garages.

The basic arrangement of aircraft in the hangar depends upon many factors. Who flies the most? That plane should be closest to the door. Which is the hardest to handle? The plane in the back of the hangar generally requires the least maneuvering. An owner of a pristine show plane might prefer to keep the aircraft in back to minimize handling.

Making them fit

Determine the ground rules with your new partner, then gather some buddies. Extra hands prevent scratches and dings during the trial-and-error period. Keep in mind, though, that they won't be around for everyday operations. All aircraft movements must be executable by a single person.

Before beginning, move both props to the horizontal position. Have two sets of chocks handy. Also, every helper should carry a short piece of two-by-four as an emergency chock. Your basic plan is to get the first airplane (the "back" plane) all

Fig. 9-11. Homebuilts are good candidates for hangar partners because they're usually smaller than factory-built airplanes.

the way into a rear corner of the hangar, then bring the other plane (the "front" plane) along the other wall.

Roll the first airplane into the hangar with the wingtip close to the left wall until the rudder is a few inches from the back. Note that the tip of the horizontal stabilizer is still 5 feet or so from the side wall. Push the tail toward that wall to swing the right wing farther back. This yaw motion is easy with a taildragger, and a nosewheel can be lifted with a little down pressure on the aft fuselage (*not the horizontal stabilizer*). You've gained a couple of inches; move the plane back some more.

Now bring the front airplane in along the opposite wall. Watch the points of closest approach; keep the two planes from swapping paint. This is where the extra help comes in handy. They should be prepared to stop the plane instantly. If the hangar door clears the front plane's spinner, you've got it made.

When they don't fit, it's usually because the left wing of the front airplane got too close to the other plane's fuselage. Reposition the front plane outside the hangar. Roll it back again, yawing to the right as necessary to delay left-wing contact. Various combinations of yaw might yield just enough space. I once kept a Citabria in a barn with a door narrower than the plane's wingspan. All it took was a see-sawing motion.

Some unnoticed characteristic of the front airplane might actually make it more suited to park in back. Try ideas as they occur; solutions seem to appear magically. Afterward, you'll shake your head and wonder why it took so long to see it.

Getting tricky

At some point, some sweaty helper might say, "Y'know, if we only had another inch or two . . .." It's time to get tricky. You've moved the planes in yaw; now try pitch and roll.

For instance, say you've rolled a high-wing taildragger into the hangar, and the trailing edge of the wing is a bit too low to let the other plane's rudder pass underneath. Lift the tailwheel of the taildragger a foot, and you'll gain 4 inches at the trailing edge. Place the tail on a stool, paint can, or box, and roll the other plane past.

Does a wingtip come too close to the top or bottom of the other plane? Try a ramp. Raising a wheel by 1 inch can drop the opposite wingtip 6 inches. Put one end of a 6-foot board atop a 4-by-4 block, and you've got a gradual ramp that'll give either wingtip at least 12 inches more clearance. It takes surprisingly little effort to pull a plane up such a shallow-angled ramp. Large-diameter or low-pressure tires easily roll onto the blunt end of a 2-inch-thick board. Otherwise, bevel the end of the ramp with a circular saw or table saw. If you use nails to hold it together, leave a hammer in the hangar for bashing any that start to work out.

Try combinations of positions, ramps, and supports as you think of them. It took 90 minutes to figure out how to put our club's Fly Baby and a Vari-Eze homebuilt into the same hangar. If the fates smile, you'll also find a combination that works.

Day-to-day operations

Take a few moments to slake the thirst of the crew and thank them. Now work out how you're going to get the planes in and out on a day-by-day basis. A good guide is a stripe on the floor showing the desired wheel path. Use paint or tape to delineate the maximum allowable excursion for one wheel. If it crosses the line, stop the plane because something's going to hit.

For added protection, make flag stands from metal or plastic tubing. Arrange the flags so the plane will tap one before it can strike an obstruction. If a flag moves, you've got a potential clearance problem.

Make sure required towbars are accessible to each owner. If a tight fit requires a nosewheel or a nonswivelable tailwheel to go sideways, set it on a trolley with casters.

Practice with help until you are used to the process solo. It becomes second nature with good guides and the right choreography. The Vari-Eze and our Fly Baby go together in seconds.

The down side

There are a few drawbacks. The obvious one is "hangar rash." Pad any areas that have tight clearances. Old couch cushions work well. You also must be willing to trust the other owner with moving your plane around. Face it, some owners aren't as careful as others.

Cockpit access can be awkward with both planes in place, so install control locks and cockpit covers before pushing the planes into the hangar. The restricted space also complicates routine maintenance. And if the hangar isn't enclosed, you might have to install additional tiedowns to match the stored positions of the aircraft.

But most folks can put up with a bit of hassle to save hundreds of dollars a year. While everyone's ideal is a nice, large hangar of their very own, waiting lists and budget-busting rents are the all-too-often reality. Find the proper partner, and your bird can sit snug and dry for little more than the cost of an outside tiedown.

AUTOFUEL BASICS

A great way to fly more cheaply is to burn autogas in your airplane. Let's look at some of the issues involved.

Is it safe to use?

Autogas has a curious reputation in the aviation world. Some pilots and mechanics swear by it; others caution against it. It probably varies with the aircraft. While I owned my 150, I had one engine failure, and it was due to contaminated *aviation* fuel.

I flew the 150 on autogas at least 50 percent of the time. Our club Fly Baby homebuilt has been operated on a near-exclusive diet of unleaded regular for the past 8 years without a hitch.

There are a couple of potential problem areas for new autogas users. First, some carburetors have floats made of composite materials. These materials can sometimes absorb autogas, which "waterlogs" the float and causes a too-rich condition.

Second, some carburetors have used neoprene-tipped needle valves; these also have shown some susceptibility to variation in the fuel. The same is true of other rubber parts in the fuel flow (such as those in the gascolator) and the varnish on old-fashioned cork-type fuel floats.

The interesting thing about these problems is that they can *also* be caused by exposure to 100LL aviation fuel! So if your plane has been burning 100LL, the carb and fuel system are probably OK. In any case, these problems can be corrected without major disassembly or rebuild. See your A&P.

Homebuilts often use special compounds to coat and seal the inside of their fuel tanks. Occasionally this material can be attacked by some components of autofuel. Check with the sealant manufacturer; monitor the fuel strainer for signs that the material might be deteriorating.

Finally, some aircraft types just seem to tolerate autogas better than others. Talk with other owners and check with the type clubs.

Some pilots who fly aircraft with multiple fuel tanks hedge their bets: They fill one tank with autogas and the other tank with aviation fuel. They take off and land on the avgas, and cruise on car gas. That sounds like a nice compromise.

Autogas has one weird yet harmless side effect. During cold weather, atomized gasoline has a tendency to condense on the inside of the intake manifold. It doesn't affect operations, but the recondensed gas tends to dribble downhill and drip out of the carburetor. It happens if the engine stops after running for just a short while. Don't sweat it unless the gas seems to keep flowing; it might then be a stuck carb float.

Getting approval

To run autogas in your airplane, you'll have to purchase an STC from either EAA or Peterson Aviation. To be eligible, both the airframe and powerplant must receive an STC. Why both? Simple: The airframe must be approved for storing and delivering fuel, and the engine must be approved to burn it.

Generally speaking, if an aircraft was originally certified to run on 80-octane aviation fuel, both EAA and Peterson have autofuel STCs available for it. In addition, Peterson has some STCs for aircraft certified for higher octane fuel that usually require some sort of modifications. Do your research, and carefully calculate if the modifications would be economically worthwhile for your airplane operations.

Either vendor will welcome your research questions and subsequent order. Contact:

EAA Flight Research Center
1145 W. 20th Ave.
Oshkosh, WI 54901
414-426-4843

Or:

Peterson Aviation, Inc.
Rt. 1 Box 18
Minden, NE 68959
308-832-2050

They'll be happy to take your credit-card order over the phone. In addition to your card number, have the following information ready:

- Aircraft make and model
- Aircraft serial number
- Aircraft registration number (N number)
- Engine make and model
- Engine serial number
- Number of fuel caps

They'll send you a copy of the STC, plus a set of labels to replace the ones by the fuel tank fillers. Your old ones say something like, "Service this aircraft with 80-octane aviation gasoline." The new ones are similar, but add: "or 87-octane automobile fuel conforming to ASTM D-4814."

Take the paperwork and the labels to your A&P. He or she will execute an FAA Form 337, apply the labels, and make the appropriate entry into your airframe and engine logbooks. As of that point, you're free to use autogas.

"That's it?" Some of you might be wondering. "Why not just use the stuff and *not buy the STC?"*

First, FBOs who carry autogas won't sell it to you unless the plane has the official stickers by the fuel caps. Second, it's illegal. If an FAA inspector walks by while you're fueling and asks to see your STC, you're hosed. Third, it might negate any hull insurance that you might carry. If you have an accident that is caused by an engine failure and the NTSB finds autofuel in your tanks, the insurer might be well within his or her rights to deny coverage if you don't have a legally executed STC.

The right stuff

If you look at the fine print of your STC, you'll see a line requiring you to use gasoline that meets ASTM (American Society of Testing Materials) standards D-439 or D-4814. Most states require that autogas conform to this standard. As of this writing, the following states do not: Alaska, Kentucky, Nebraska, New Hampshire, Ohio, Oregon, Pennsylvania, and West Virginia. But even if your state doesn't *require* compliance, the fuel sold there probably follows the standard anyway.

Autogas changes with the season. The refiners alter the formula slightly in the winter, for instance, to aid starting. They change it again during the summer to inhibit vapor lock. It's important to buy the correct gas because you don't want to run "winter" gas in the summer!

How do you tell? You can't, really, but you can decrease your risk by buying your autogas from a major-brand retailer. A retailer down the street might have it for 5 cents a gallon cheaper, but the station might operate by buying up leftover "winter" gas in the springtime.

Also, buy your gas from a popular location because a place that sells a lot of gas is resupplied more often, and the gas will be fresher.

Many states and metropolitan areas require the use of "oxygenated" fuels to reduce pollution. Refineries oxygenate the fuel two ways: adding MTBE (methyl tertiary butyl ether)—I had to ask my chemist wife for this one—or adding ethanol, a type of alcohol.

If MTBE is used in your area, no problem. The FAA has determined that it doesn't cause any problems and has OK'd its use in aircraft holding autofuel STCs.

Ethanol is different. Some aircraft fuel systems can't stand the stuff. Its usual modus operandi is to cause the swelling of any rubber compound: gaskets, seals, needle tips, and the like. It has a greater tendency for vapor lock, and it can cause engine damage from excessive combustion temperatures.

Automakers understand that fuel quality can vary and design their systems accordingly. Cars are designed to be insensitive to alcohol in the fuel, but you *cannot put gas with ethanol in your fuel tanks under the terms of your STC*. How do you tell?

Ethanol testing

Fortunately, testing for ethanol is simple. As a type of alcohol, it has a great affinity for water. You'll need a small, clear container. A test tube or gradated cylinder will do; you can often buy them at hobby shops (look for the chemistry display) or photo supply stores. I use a small jar. Whatever kind of container you use, it must come with a stopper or lid to adequately seal the top.

Take your container and fill it 10 percent full of tap water. If you use a gradated cylinder, great; just remember (or mark) which line you filled to. On unmarked containers, put a line on the outside of the jar to mark the top of the water. Use a crayon, grease pencil, or transparency marker.

Fill the container with the fuel to be tested. You'll notice that the demarcation point between fuel and water is readily visible. Cap the container, shake thoroughly, then let it stand at least three minutes or until the fuel looks normal again.

Check the demarcation point. If it's the same as before, great; there's no alcohol in the fuel. If there is apparently more water than at the beginning, ethanol is in the fuel. *Don't put it in your airplane!*

Alcohol combines with water a lot more readily than gasoline; in your sample container, ethanol has separated from the gas and joined with the water. This testing procedure is summarized in Fig. 9-12.

You can also buy a volatility tester to determine the fuel's susceptibility to vapor lock. It's available through Peterson Aviation for about $50.

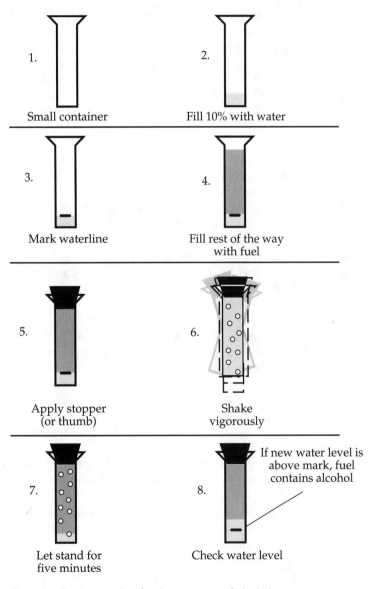

1. Small container
2. Fill 10% with water
3. Mark waterline
4. Fill rest of the way with fuel
5. Apply stopper (or thumb)
6. Shake vigorously
7. Let stand for five minutes
8. Check water level

If new water level is above mark, fuel contains alcohol

Fig. 9-12. Testing autofuel for the presence of alcohol.

Logistics of autogas

By far the biggest problem with using autofuel in aircraft is the logistics. If a local FBO doesn't sell car gas, you'll have to haul it yourself from a local gas station. Then you've got to get it out of whatever containers you have and into your aircraft.

The type of container depends upon how much gas you think you'll use. I used a 5-gallon and a 2-gallon can for fueling my 150. Most of my flights were just for local sightseeing or touch-and-goes; therefore, 7 gallons worked well. Besides, the gas station was right across the street.

Five-gallon cans are the largest size that one person can conveniently handle. Transporting any more than two or three cans gets unwieldy. Several folks at my airport use larger containers that are semipermanently installed in pickup trucks. That is a good solution for flying, but it does limit the utility of the vehicle. Some full-size pickups have separate saddle tanks. Those might be a good possibility.

Most people seem to get by with a couple of 5-gallon cans. Get metal or plastic containers that are approved for carrying gasoline. Ordinary plastic containers that are not approved can build up a static-electricity charge.

Another thing you'll need is a *grounding wire*. Obtain 10–15 feet of 18-gauge or bigger stranded copper wire. Attach a large alligator clip at both ends. Before starting fueling, clip one end to the gas can. Then attach the other end to a well-grounded point on the aircraft to equalize the static charge between the can and the plane.

Connect them in the above sequence. *Don't do it the other way around.* A spark might jump; it's a lot better if it jumps from the wire to the airplane than to a full can of gas.

As far as transferring fuel, getting the gas into the cans is easy enough. Getting it into the aircraft can be done with gravity or some sort of pump. I use gravity, but I have never really liked it. I put a funnel in the tank filler, then hoist the 5-gallon can atop my shoulder to pour gas into the tank (Fig. 9-13).

It's not all that easy. When full, the can weighs more than 30 pounds. It's unsteady, too. With my current plane, I can do it while standing on the ground. My 150 required a stepladder. I was uncomfortable hauling 30 pounds of gasoline 8 feet into the air. Instead, I dumped the contents of the 2-gallon can into the tank first, then refilled the smaller container from the larger 5-gallon can.

Depending on your aircraft and other aspects, you might be able to finesse the situation. One local owner places his cans atop his truck and uses a hose to his aircraft's fuel tank (Fig. 9-14). He uses a stainless steel ball valve (bought from the local hardware store) to turn the flow on and off.

Alternately, you can rig up some sort of pump. You can find gasoline handpumps through several sources. Build a plywood rack to hold the pump and a gasoline can.

A couple of folks at my home field have rigged up transfer systems using boat bilge pumps (Fig. 9-15). They empty a 5-gallon can in just a few cycles of the handle. I had some concerns about the pump's rubber diaphragms deteriorating from contact with gasoline, but it hasn't happened after a number of years. I suspect the pumps might be designed to resist fuel or oil because either can be present in a boat's bilge.

Fig. 9-13. The author pouring the last dregs of fuel into Fly Baby N500F. It is much easier than pouring directly into wing tanks of a high-wing airplane.

Fig. 9-14. A simpler solution: Bill Ashby uses gravity to fill his Aeronca's tank.

The ultimate solution is an electric transfer pump. These sell for about $200; you'll want the kind that run on 12 Vdc so you don't have to run an extension cord out to your tiedown spot. They even include genuine fuel nozzles, the kind that shut off the flow when the tank is full.

I shouldn't have to caution you about trying to rig up your own electric pump because the dangers of sparks and fuel vapors should be obvious. One bright idea some people get is to use compressed air to transfer the gas. One problem: It actu-

Fig. 9-15. Homemade fuel transfer system using a boat bilge pump.

ally *increases* the danger. Compressing the tank raises the fuel-air mixture closer to optimal, just like inside an engine cylinder during compression.

When shopping for pumps, consider the transfer rate as well. The local hardware store sells little hand-operated pumps that might look good, but they only have a ¼" tube. It would take a long time to move 30 gallons of gas. A ¾" or 1" hose would be better.

Besides transferring the fuel, you should rig up some way to filter it to prevent dirt or water from entering your plane's tanks. If you're pouring gas by hand, you'll need a funnel anyway. Aviation suppliers sell funnels specifically for fueling aircraft; these include filters to eliminate dirt and even water.

With a pump arrangement, you could rig up an in-line filter. These can be found at larger hardware stores or farm stores. The filters are very much like the gascolator in your aircraft, designed to separate water and dirt.

FINAL FUELING WARNING

Whether you use aviation or automotive fuel in your aircraft, at some point you're probably going to be pumping it yourself. While it seems no different from putting

Problem: Nozzle causes stress on filler pipe

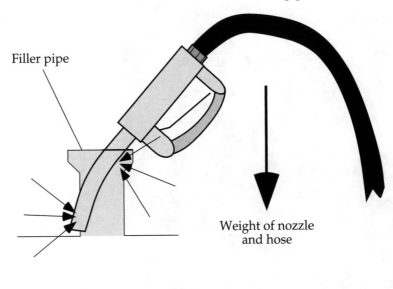

Filler pipe

Weight of nozzle
and hose

Solution: Avoid solid contact with pipe wall

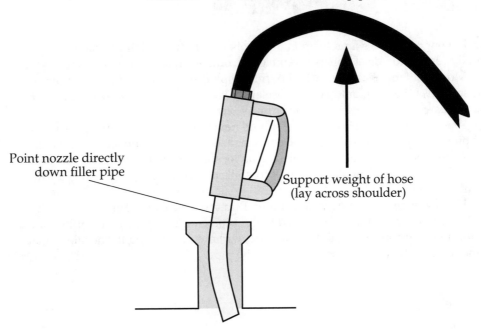

Point nozzle directly
down filler pipe

Support weight of hose
(lay across shoulder)

Fig. 9-16. Don't damage your fuel tanks while filling!

Fig. 9-17. Drape the hose over your shoulder to keep from damaging the filler neck. The truck contains a 200-gallon tank and electric transfer pump.

gas in your car, there's a problem many pilots aren't aware of. Incorrect technique could cause hundreds of dollars in damage; it could even bring your airplane down.

Most fuel tanks have a short bit of filler pipe running from the tank proper to the surface of the wing or top of the fuselage. *This filler pipe is not designed to support the weight of the nozzle and hose of the fueling system.*

As shown in Fig. 9-16, the nozzle should never be allowed to apply pressure to the pipe. The typical mistake is to place the nozzle in the pipe and allow the weight of the hose to hold the nozzle in place. The localized pressure this generates can crack the pipe, possibly allowing fuel to spill into your cockpit or wing during flight.

Don't wedge the nozzle into the tank. Run the hose over your shoulder (Fig. 9-17) so the weight cannot be applied to the pipe. You will provide complete and continuous support for the hose and nozzle, without positioning the nozzle so that it independently maintains its position in the pipe.

10

Keeping it legal

DID YOU REALIZE that you were risking an FAR violation every time you rented an aircraft? Take a look at FAR Part 91, Subpart E: Maintenance, Preventative Maintenance, and Alterations. Note how nearly every section includes either "The owner or operator of an aircraft . . ." or "No person may operate an aircraft . . .."

In other words, if the FAA ramp-checked you and the aircraft's annual inspection was overdue, you would take the rap. Did you check the aircraft's maintenance logs before each flight? Did you ensure that all replaced parts were legal? Did you verify that any modifications had been done in accordance with the regulations?

Probably not. Yet as the pilot, the responsibility was yours. Now that you own your own plane, the risk is greatly decreased; however, the responsibility is still there. Now, of course, the full force rests on you.

INSPECTIONS

What inspections are required for your aircraft to be legal? There are four primary types:

- 100-hour inspections
- Instrument/equipment calibration/inspections
- AD-required inspections
- Annual inspections

The annual is a major topic that needs an entire subsection to itself. For now, let's consider the other three.

100-hour inspections

As discussed in chapter 2, 100-hour inspections are required *only for aircraft operated for hire* (Fig. 10-1). They won't apply to you as an owner unless you operate the plane commercially or have it on leaseback to the local FBO.

Fig. 10-1. This Beech A36 Bonanza only requires an annual inspection if operated privately; however, 100-hour inspections are necessary if on leaseback or otherwise flown commercially.

It's a good thing that the 100-hour is not required for a private owner from the cost point of view. There's little difference between an annual and a 100-hour, as defined in FAR 43 Appendix D. The only advantage a 100-hour has is that it can be performed by an ordinary A&P; the inspection authorization rating is not required.

Even though it isn't required, some feel that a 100-hour should be done, just as a matter of safety. The decision is yours. Unless you fly 200 hours or more a year, I wouldn't even consider it. Even then, it would depend on the complexity, history, and general reliability of your aircraft.

Instrument/equipment inspections

Some equipment needs separate inspections and checks at prescribed intervals. FAR 91.413 requires that the aircraft's transponder undergo test and inspection every 24 calendar months. If you fly IFR, FAR 91.411 requires an inspection and check of the altimeter and static system at the same interval.

It would be most convenient to schedule these tests in conjunction with an annual inspection; however, many mechanics are not authorized to perform some tests that are reserved for specialists at certified repair stations. If your mechanic is located at the same airport as one of these outfits, the tests could easily be combined.

Airworthiness directives

Airworthiness directives can be either *one-time* or *recurring*. One-time ADs typically specify that a particular component be inspected, modified, or replaced.

The AD will specify how soon the operation must be performed, generally based on flight time, for instance, "Within the next 50 flight hours." Sometimes the time interval is sufficient to let you schedule the AD work in conjunction with the next annual.

Recurring airworthiness directives specify that a special inspection must be performed at given intervals. These range from fairly short (which can usually be canceled by the replacement of the offending component with an improved one) to several hundred hours. Similarly, these inspections might be relatively simple, or might require the use of special techniques such as dye penetrant and ultraviolet light. The existence of such ADs is one of the things you should have determined during your shopping phase.

Normally, the inspection interval is well-enough spaced that it can be combined with an annual. It's easy enough to predict if the aircraft's flight time will cross the magic number in the coming year.

Logbook entries

It's not enough to *do* the inspections, though. They must be entered properly in the appropriate logbooks. It's the responsibility of the mechanic to make the log entries, but it's in your best interest to ensure the entries are as precise and accurate as possible.

At a minimum, the person performing the inspection or procedure must include the type of procedure performed, the date, the mechanics' signature, and his or her certificate number. More information is necessary in most cases. For the altimeter/static system inspection, the maximum altitude to which the test was performed should be included.

For preventative maintenance, enough details should be provided to ensure that anyone reading the log 10 years hence will understand exactly what was done. "Rebuilt right brake" is insufficient. You would prefer that the log read something like "Honed right brake cylinder, replaced O-rings part #XXX-XXX and left and right brake pads, P/N XXXX-XXX. Replenished system with MIL-H-5606BB brake fluid." And again, the mechanic should date, sign, and include his or her certificate number.

The finer details don't have to be included if there's a manufacturer's standard procedure to reference: "Rebuilt right brake in accordance with Service Bulletin 82-105."

For ADs, precision is even more important because you might be selling the aircraft in five years. Some bright pilot (who undoubtedly read chapter 7 of this book) might look through your logs and say, "You didn't perform AD 93-10-4." You'll point and say, "Here it is, 'Brake AD complied with.'"

But of course, nothing says *that* particular AD was performed. At the minimum, the log entry should include the AD number and issuance date, the number of hours on the aircraft/component at the time of compliance, how the AD was complied with (i.e., inspection, replacement, you may directly reference the AD's instructions), and when the procedure will next be required. The entry must be wrapped up with the date of compliance and the signature/certificate number of the mechanic.

The FAA also requires that a number of other factors be tracked and recorded. They don't necessarily have to be kept in the aircraft logbooks, but must be permanently retained. Might as well put 'em in the logs as anywhere else.

FAR 91.417 lists the items that must be recorded. In addition to the data on inspections, calibrations, and AD compliance, you must track:

- The total time in service of the airframe, engine, and propeller.
- The current status of life-limited parts of each airframe, engine, and propeller.
- The time since last overhaul of all items installed on the aircraft that are required to be overhauled on a specified-time basis.

Protect those logbooks!

From the preceding, it's obvious you'd be in a world of hurt if the airframe, engine, or propeller logs disappear. As mentioned in chapter 6, a no-logs airplane is in a sticky situation. For instance, you'll have no proof of engine or airframe time nor records of compliance with any AD.

I kept my 150's logs in the aircraft; I'm smarter now. Contrary to what some folks say, the FAA does *not* require the logbooks to be carried aboard the airplane. You are required to present them only upon reasonable notice. So find a safe place for them. Photocopies stored in a separate location provide additional peace of mind.

THE ANNUAL

To be legal to fly, your aircraft must have undergone an annual inspection in accordance with FAR 43 within the past 12 calendar months. What constitutes an annual?

Basically, the inspector is required to inspect the aircraft in order to determine whether it meets all applicable airworthiness requirements (Fig. 10-2). He or she must use a checklist that has been compiled by the inspector or another source, such as the aircraft manufacturer. Then the aircraft engine must be run to determine if its performance is satisfactory. The inspector is required to check the static and idle RPM, the magnetos, the fuel and oil pressure, and the oil and cylinder temperature.

WHO PERFORMS IT?

By FAR 65.95, only an A&P holding an inspection authorization (IA) is allowed to perform an annual. To gain an IA rating, an applicant must have held an A&P license for at least three years, must have been "actively engaged" in maintaining aircraft for at least the past two years, must have the equipment and facilities available to perform the tasks of an IA, and must pass a written examination.

In addition to performing annual inspections, an IA is authorized to inspect and return to service aircraft that have undergone major repairs or modifications and supervise *progressive inspections*, which are a special form of annual.

The IA rating expires on March 31 and must be renewed every year. At renewal time, the applicant must prove that he or she still meets the qualifications

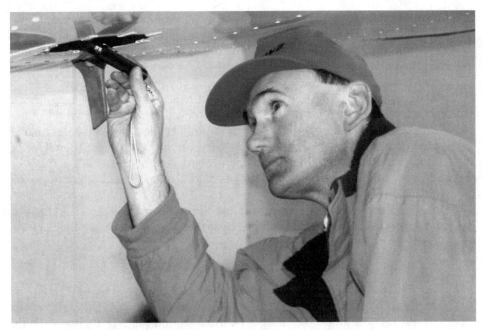

Fig. 10-2. The annual is an in-depth analysis of the aircraft's condition.

for gaining the license and has attended a refresher course in the previous year. In addition, he or she must have been actively performing the inspections that require an IA rating and pass an oral exam by an FAA inspector on their knowledge of FAA regulations and workmanship standards.

At some facilities, your aircraft will be inspected by an ordinary A&P. He or she will find the trouble spots and correct them. Then an IA will be called in, inspect the aircraft, and sign the logs.

Selecting the shop

Where should you take your plane for the annual? In the first place, the shop should be familiar with the type of aircraft that you own. If you've bought a Cherokee or a 172, you shouldn't have any problems. With a Champ or Stinson, you'll want to find an outfit familiar with the foibles of these older birds. It makes no sense to pay someone to learn how to work on your type of plane. Any number of A&Ps haven't done fabric repairs since their student days.

Some mechanics, especially the smaller independents, might even refuse to work on certain types. "I won't touch the ––––––," one said to me. Apparently, the inspection panels for this model were all custom-made because they only go on one spot, in one particular orientation.

In my opinion, the best solution is to get recommendations from fellow airplane owners. Join local pilot's groups or EAA chapters and ask around. You'll get a variety of opinions.

Of course, no one is perfect or unbiased. I was real happy with the man who did my 150's inspections. He was thorough and recognized that I didn't have a lot of money for nonrequired repairs. Imagine my surprise, then, when I saw a note on a local airport's bulletin board warning *against* my mechanic. Didn't make any difference to me.

Make no mistake about it, picking the right mechanic *can* make a significant difference to your pocketbook. Like all businesses, there are those who are less than ethical. Even worse, they can impart significant leverage that the aircraft owner can find hard to fight. If they feel certain repairs are required, they'll refuse to sign off your annual until the work is performed. And because your plane is lying half-disassembled on their shop floor, it's hard to take it to a different mechanic for even a quote.

So chose carefully *before* turning your aircraft over. And remember, sometimes aircraft *do* require repairs. If the mechanic does find a problem, take a look at it before starting to point a finger.

Cost

Talk about the anticipated cost prior to selecting the shop. Aircraft service manuals indicate the amount of time an annual should take. That should be the starting point.

The service manual times vary by aircraft type. The Cessna 150 manual estimates 13 hours for an annual, which is actually more than mine took. A 172 takes about 16 hours, and a retractable might take 20 hours or more.

Typical shop rates are quite reasonable, really. I've had my airplane worked on for a lower per-hour rate than the local General Motors dealership charged. Still, the annual on that 172 might cost $800 with a typical $50-per-hour shop rate.

Some shops operate on a *flat-rate* basis, which means that the basic charge for the annual is based on the time predicted in the service manual multiplied by the shop's labor rate. This price includes all "normal" activities and typically includes most minor repairs. Parts are usually extra.

There are pluses and minuses to flat-rate annuals. It does allow you to predict more closely what the annual will cost. Most problems found during annuals are pretty minor. These little problems can add up with a shop that charges by the actual time worked; it wouldn't change the price of a flat-rate operation. There are obvious potentials for abuse if the annual charge is not capped somehow.

On the other hand, the shop performing a flat-rate annual loses money if your plane takes longer than the shop manual's recommended time. There might be a tendency to cut corners; little problems might be ignored. Plus, make sure you understand *exactly* what is covered under the flat rate. The threat posed by an unscrupulous operator is obvious.

One way to reduce the cost of the annual is by participating in it. Many shops have reduced rates if the owner does the "grunt" work: wash the plane and engine, undo the cowling and inspection panels, remove the interior, and the like. One flat-rate shop I talked to said they reduced the time by three hours if the owner participated.

Participating in the annual is not for everyone, of course. A Bonanza annual might take three days to a week; that's a lot of time off for a $150 saving. If your time is shorter than your money, there's a lot to be said for simply turning the airplane over to the shop.

If you decide to participate, find out the ground rules in advance. When will they actually need you? It might be only at the beginning and the end if all you'll be doing is cleaning and reassembling the plane. What tools should you bring? Most professional mechanics prefer not to have let others rifle through their toolboxes. You might be asked to bring your own screwdrivers and wrenches.

Finally, if there's other work that you would like performed on the aircraft, the annual is the time to do it. After all, the plane has to be opened up *anyway*. Perhaps there have been some slight problems, or maybe you'd like to have some changes made (Fig. 10-3). Make sure you discuss it with the mechanic, and make sure you understand how much the additional work will cost.

Fig. 10-3. The annual is a good time to get small glitches fixed. An electronic thermometer and probe can be used to check out an oil temperature problem.

Scheduling

A nice thing about the annual is that it's predictable. You know that the current annual will expire 12 calendar months after the sign-off. If the annual was signed off on July 10, 1997, it's good through July 31, 1998.

However, there is room for a bit of acceptable and legal trickery. The applicable FAR (91.409) doesn't require a yearly annual; it requires that one be performed *within the past year*. But say you live in a colder climate where you won't fly through the winter. Assume that your new plane's annual is due in December.

There is nothing wrong with leaving your first annual until the spring thaw. Indeed, having a mechanic pore over your plane at the start of the flying season rather than at the end would probably be safer all around.

You can also use the "calendar-year" provision to gradually shift your annual date to a more convenient time. The annual for our club's Fly Baby used to be due in June; however, that was fairly inconvenient. The summer months are the busiest flying time, especially for an open-cockpit airplane! We didn't want the plane laid up for a week during this period.

So we started the annuals at the *end* of the month, timing the first one so that the mechanic's signoff was dated the first of July. That made it valid until the end of the following July. We repeated the process for two more years until the annual was due in September. Our flying slackens but doesn't die off during the winter. When spring comes around, the engine is due for an oil change, which gives us a good excuse to pull the cowling off and perform our own safety inspection for the coming summer.

Delivery

When delivering the aircraft to the shop, it's important to be as ready as possible. One thing the IA will check for is if all the required documentation is aboard the aircraft. Make sure that the airworthiness certificate, registration, radio license, weight and balance, and operations manual are there.

Airframe, engine, and propeller logbooks are also needed, not just to make the final notation, but the inspector has to verify that the ADs have been complied with and determine the total time in service of various components.

Provide telephone numbers where you can be reached during the annual, in case something turns up and you have to be contacted.

Finally, if you're having other work done, include a written list of squawks and problems. Be as specific as possible. Tell exactly what the conditions are when the problems occur.

Functions to be performed

As mentioned earlier, the IA must use a checklist to guide the inspection process. The checklist, no matter what its source, probably began with the guidance found in FAR Part 43, Appendix D.

FAR 43 requires that all inspection plates, access doors, fairings, and cowlings be removed, and the engine and airframe thoroughly cleaned (Fig. 10-4). A whole smorgasbord of inspections follow: fabric and skin, cabin and cockpit, engine and nacelle, landing gear, wing and center section, propeller, controls, and radio (Fig. 10-5).

Fig. 10-4. A thorough cleaning of the engine and the airframe are part of any annual.

Of course, FAR 43 breaks down the inspections to far greater detail. Here's what it says for just one section:

(d) Each person performing an annual or 100-hour inspection shall inspect (where applicable) components of the engine and nacelle group as follows:

(1) Engine section—for visual evidence of excessive oil, fuel, or hydraulic leaks, and sources of such leaks.

(2) Studs and nuts—for improper torquing and obvious defects.

(3) Internal engine—for cylinder compression and for metal particles or foreign matter on screens and sump drain plugs. If there is weak cylinder compression, check for improper internal tolerances.

(4) Engine mount—for cracks, looseness of mounting, and looseness of engine to mount.

(5) Flexible vibration dampeners—for poor condition and deterioration.

(6) Engine controls—for defects, improper travel, and improper safetying.

(7) Lines, hoses, and clamps—for leaks, improper condition and looseness

(8) Exhaust stacks—for cracks, defects, and improper attachment.

(9) Accessories—for apparent defects in security of mounting.

(10) All systems—for improper installation, poor general condition, defects, and insecure attachment.

(11) Cowling—for cracks and defects.

That is some list, and this is just for the engine! In addition to the FAR 43 tasks, the aircraft service manual identifies specific operations that must be performed at

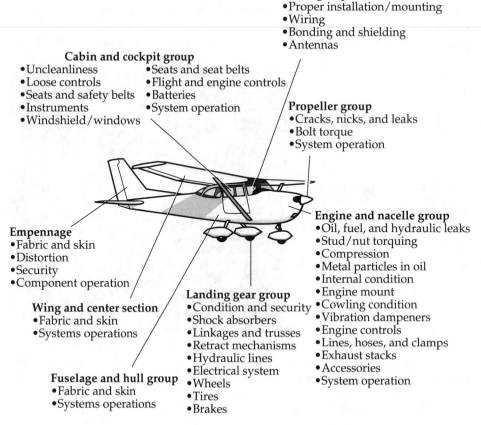

Radio group
- Proper installation/mounting
- Wiring
- Bonding and shielding
- Antennas

Cabin and cockpit group
- Uncleanliness
- Loose controls
- Seats and safety belts
- Instruments
- Windshield/windows
- Seats and seat belts
- Flight and engine controls
- Batteries
- System operation

Propeller group
- Cracks, nicks, and leaks
- Bolt torque
- System operation

Empennage
- Fabric and skin
- Distortion
- Security
- Component operation

Wing and center section
- Fabric and skin
- Systems operations

Fuselage and hull group
- Fabric and skin
- Systems operations

Landing gear group
- Condition and security
- Shock absorbers
- Linkages and trusses
- Retract mechanisms
- Hydraulic lines
- Electrical system
- Wheels
- Tires
- Brakes

Engine and nacelle group
- Oil, fuel, and hydraulic leaks
- Stud/nut torquing
- Compression
- Metal particles in oil
- Internal condition
- Engine mount
- Cowling condition
- Vibration dampeners
- Engine controls
- Lines, hoses, and clamps
- Exhaust stacks
- Accessories
- System operation

Fig. 10-5. Annual inspection target areas.

each annual. The service manual will specify lubrication points, for example, and might provide specific details for how to inspect certain parts.

The mechanic will either make a list of problems to be fixed and continue the inspection or will fix each problem as it is discovered.

Completion

Eventually, it comes time to pay the—gulp—bill and pick up the aircraft. Check the logbooks and make sure the IA has made the required entry. It should be similar to the entries your instructor made in your pilot's log: "I certify that this aircraft has undergone an annual inspection on July 10, 1996, and is airworthy." The IA should sign it and include his or her certificate number. The same entry should be made *in all logbooks*. Check!

If any ADs were complied with, the appropriate log entry should also have been made: "Oil filter modifications made in accordance to Section 2 of AD-75-01." The entry must be completed with a signature and certificate number, of course.

After settling up with the shop, don't hop in and fly just yet. Sad to say, even the best mechanics sometimes forget something. It's crucial to your safety and the well-being of the aircraft to perform a very thorough preflight after any sort of maintenance.

Don't head right out on a cross-country, either. Stay in the pattern. Do a couple of touch-and-goes. Land, taxi back to your tiedown spot, and check things over *again*. Sigh contentedly, and wipe the sweat off your brow. It's over for another year.

REPLACEMENT PARTS

One thing you'll really cringe at is the price of replacement parts for your bird. I'm not going to try to justify the manufacturer's pricing schemes. There are some good reasons why aircraft parts cost more than the equivalent hardware-store items.

The type certificate

To prove the original design's safety, the manufacturer performed exhaustive structural analyses. Aircraft were ripped apart in ground test fixtures. Prototypes generated reams of flight test data. When the FAA was convinced that the aircraft met its standards, the design was granted a *type certificate*, or TC. The TC actually defines the aircraft, describing the components down to the last bolt.

The type certificate isn't a mere bureaucratic ticket to be punched on the way to production. If the aircraft conforms to its TC, it automatically meets the same standards as the prototype. The TC is the pilot's guarantee: Barring improper maintenance, any particular aircraft will perform and handle like any other of the same model. Unless an aircraft conforms exactly to the components and construction shown on its type certificate, it cannot be legally flown.

Herein comes the rub. The TC specifies which part numbers by which manufacturers are installed where. If you replace any part with one not specified in the type certificate, you've just busted the regs.

Legal replacements

What kind of parts *can* you install, then? Obviously you can use parts built by the component's original manufacturer. Called *original equipment manufacture* (OEM) parts, they are identical to what should already be installed in your aircraft. They can be obtained new through a parts supplier. Used parts are an option.

Aircraft salvagers are good sources for used parts, especially for those little trivial components that the manufacturers seem to delight in overcharging for. My old 150 had a pull-cable starter. The starter gear was engaged manually when the pilot pulled on a handle in the cockpit. It was a simple system that was not subjected to the starter woes of later models of the O-200.

At least it seemed simple to me until both "ears" broke off the handle. I needed a new cable, and Cessna wanted almost $100 for the part! The local auto-parts store had a similar unit for $10, but it wasn't approved, and who wanted a handle labeled "Choke" in their airplane?

I called a few aircraft salvagers. One had the cable, and it cost $30 plus shipping.

Components such as magnetos and starters are sold with a little paperwork involved. You'll hear the word *"tagged"*: yellow-tagged, green-tagged, and red-tagged. The term means that the used component's condition has been formally evaluated by a certified repair station. *Yellow* means the part is serviceable and airworthy. *Green* indicates the part isn't currently airworthy, but is repairable. Units in real bad shape earn *red* tags—fit only for scrap.

The parts that you get from the salvager probably won't have a tag. In this case, you and your mechanic take the responsibility for using the component. As of this writing, it's legal. Unfortunately, bogus aviation parts have been causing problems lately. The FAA is considering a requirement that all parts installed on an aircraft must have a traceable heritage. That would essentially require that all installed parts have yellow tags. If you use a yellow-tagged part, retain the tag in the aircraft logbooks.

Parts for your aircraft *can* be made by someone other than the original manufacturer. Other companies can earn *parts manufacturer approval* (PMA) from the FAA by proving that their parts meet the same requirements as the OEM parts. They can do this by making their parts identical to OEM parts, by gaining a license from the OEM manufacturer, or by performing the engineering required to prove the part is as good or better than the original. Such parts are marked "FAA-PMA" along with information regarding where the part is authorized for use.

Hardware

One small exception to OEM/PMA parts is small hardware. The U.S. government developed standards for the construction and testing of aircraft-grade hardware. The aircraft and component manufacturers will specify nuts, bolts, washers, and other parts based on one of the following specifications:

- MS (military standard)
- NAS (national aerospace standard)
- AN (Air Force/Navy)

These specifications will list the minimum and, in some cases, the maximum requirements for approval. These requirements include dimensions, tolerances, strengths, and finishes.

Most of the hardware used on your aircraft will probably be of the AN variety. This is the granddaddy of standards; in fact, the AN used to stand for Army-Navy. "What's so special about the AN standard? Why can't I just buy bolts at the hardware store?" For one thing, the ordinary bolts from the hardware store aren't very strong. The most common units are about a third the strength as the same-size AN bolts.

Second, AN specifications usually require that parts be cadmium-plated. When steel comes in direct contact with aluminum, a little moisture can trigger a phenomenon called *dissimilar metals corrosion*. The cadmium plating of an AN bolt prevents its steel from coming in direct contact with whatever it is holding together.

Finally, the most important thing about the AN standards is they are just that—*standards*. The bolt maker certifies that *every* unit meets the requirements of the specification. Maybe the bolt in the hardware-store bin this week is just as strong, but the bolt in the same spot next week might not be as strong.

Your mechanic will replace hardware in your plane with parts meeting AN/NAS/MS standards. Make sure you do the same.

MODIFICATIONS

There isn't a pilot in the world completely satisfied with his or her airplane. Some deficiencies are fondly regarded as quirks, mere beauty marks on their beloved's cheek. Serious problems usually result in a change of ownership.

Between these two extremes, the decision regarding what to accept is harder. Sure you like the plane, except, well, it doesn't have good short-field performance (Fig. 10-6), or it won't fly far enough, or the FAA says you have to install a new black box.

Fig. 10-6. A 65-hp standard Chief (right) meets a 115-hp Super Chief. The speed increase of this STC'd modification pales in comparison to the plane's new short-field ability.

But the government is real persnickety about aircraft modifications. The FAA is perfectly willing to approve almost any change that won't affect the reliability or safety of the aircraft; however, the burden of proof is on the modifier to prove the altered aircraft meets the same safety requirements as the original design.

We've already covered the process that the manufacturer went through to get a type certificate for your aircraft. Recall that if an airplane doesn't conform exactly to the components and construction shown on the type certificate, that airplane's airworthiness certificate is invalid.

However, the situation isn't totally rigid. The FAA allows minor modifications under the *field approval process*, using the FAA Form 337 (explained in the next subsection). A wide range of changes may be made, but the alterations cannot affect the aircraft's performance, handling, or flight characteristics. If they do, or in the opinion of the FAA there is a chance that they might, the modifications must be thoroughly analyzed to support the issuance of an addendum to the original type certificate. The addendum is a *supplemental type certificate* (STC). Let's examine Form 337 mods first.

Field-approval modifications

The most common field approval is for avionics installation. The Form 337 can be executed by any A&P with an IA authorization. The FAA Flight Standards Maintenance Branch must eventually endorse the completed paperwork, but the aircraft can be returned to service when the IA signs the form. That way, you aren't grounded waiting approval for a routine radio installation.

However, the flight standards branch is not a rubber-stamp organization. If they feel performance, handling, or flight characteristics will be affected, field approval will be denied. If a change is unusual or the IA has any doubts, the IA can refuse to return the aircraft to service until the FAA has actually signed the paperwork.

Your best move is to obtain the flight standard branch's approval before the modification is made. One big difference between a field approval and an STC is that the STC process might require inspection and documentation of the work in progress. If you've already made the changes, and the FAA decides an STC is required, you'll have to open the aircraft up enough to allow the internal parts to be inspected.

A week's work of research and discussion before the change is started can avoid months of argument and downtime. Flight standard's airworthiness inspectors are reasonable. If you can find a similar mod performed under a Form 337, they'll probably sign off on yours. One pilot bought a Cessna 195 in Canada and brought it into the United States; however, it had a different engine installation, and his local flight standards office refused to grant a field approval. The owner talked with inspectors at another region's flight standards offices and found an inspector who had approved the same configuration. After a short three-way conversation, the local office approved the change.

Field approval is a relatively inexpensive process. Other than the actual costs to alter the aircraft, the main expense is the IA's time to recalculate weight and balance and to describe the modification on the back of Form 337. Laying careful groundwork with flight standards beforehand will smooth FAA approval.

However, if they won't accept the change as a field modification, an STC is the only option. You're facing some pretty involved engineering analyses and structural tests. You'll probably need some hired guns, and they're not cheap.

Buying an existing STC

Before starting a range war with the FAA, perhaps there's an easier way. Is the same modification sold in the form of an approved multiple STC? The FAA maintains a summary of available STCs. The summary supplies a brief description of each modification and the address of the STC holder. It's a mighty big document, two thick volumes of fine print. The section on Cessna 210s alone is more than 11 pages.

Most owners find out about available modifications through advertisements; however, there are a few things to watch for. First, the modification must be FAA-approved under a multiple STC. Beware of the terms "fits" or "adaptable to." Just because something can be bolted onto an aircraft doesn't mean the FAA has OK'd it.

If the STC holder sells components to support the mod, he or she must possess a PMA; otherwise the parts aren't legal. An interesting quirk in the regulations allows the STC holder to modify aircraft in the holder's facility without the need for a PMA.

No coordination with the FAA is required when performing a change in accordance with a purchased STC. The IA signs off the work and notifies the FAA via the same Form 337 used for field approvals. That's it.

The cost of buying rights to an STC varies. Most older aircraft were certified to use cotton fabric, so suppliers of more modern covering materials practically give their STCs away. If they didn't, most customers couldn't legally use their products. On a slightly higher scale, the EAA charges 75 cents a horsepower for their auto-fuel STC. The rights for some major mods can run in the thousands of dollars, not including labor or materials. It might seem like a lot of money, but not compared to the difficulty of earning your own STC.

Do-it-yourself STC

If an appropriate mod isn't on the market, you might have to plunge ahead and gain your own STC. To begin with, any part installed on a certified aircraft must be of aircraft quality. This can be proven two ways:

- The part has been manufactured under *technical standard order* (TSO) provisions.
- The modifier must prove that the part is qualified under the airworthiness FARs.

To legally install a non-TSO'd part like an automobile alternator, you must prove that the alternator meets the same requirements as the aircraft unit. If the TSO requires that the alternator generates full-rated output at 150°F while subjected to a 3G vibration, you must present test data to that effect.

"But the parts are identical!" Are they really? Sure, the cases might match, but are the field windings the same wire gauge? Does the auto unit use the same bearings as the aircraft unit? Is the built-in regulator an identical design, with identically rated components? It doesn't make much sense to risk thousands of dollars in avionics just to save a hundred bucks on an alternator, or risk even more if the alternator fails while flying in instrument conditions or at night.

Just because the part has TSO approval doesn't mean you can install it. The TSO is essentially just a bench check; it's up to you to prove that it functions properly when installed in a particular aircraft.

When you have picked out the TSO parts to be used, the first stop is at the FAA's regional aircraft certification office (ACO) to explain the proposed modification. The ACO is the approval agent for STCs. Again, talk to the FAA before you start work. The engineers of the ACO are only too happy to discuss your proposal. They'll explain the regulations and assist you during the certification process, but they aren't allowed to help design or substantiate a modification.

The ACO first determines the "certification basis" of the aircraft from the original type certificate. For the modification to receive approval, the aircraft must still meet the minimum requirements in effect at the time of its original certification.

For instance, the stall speed of an FAR 23-certified aircraft is limited to 61 knots in the landing configuration. Clip the wings as you will, but an STC cannot be granted unless the stall speed remains 61 knots or less. An older aircraft might have been certified prior to this rule and needn't comply; however, the older requirements were occasionally more strict. Planes like the Aeronca Champ must recover from a six-turn spin hands-off, while modern normal-category aircraft need only stop spinning within one turn or three seconds of the application of normal recovery techniques.

Exemptions can be made. An example would be a turbine engine installed on an aircraft built before turbines were certified. The ACO would use more modern regulations where the engine is concerned. Compromises between older and newer regulations are allowed, but it's up to the ACO.

The regulations specified in the certification basis identify those factors that must be proven by engineering analysis or ground testing versus those that require flight testing. The ACO will determine the amount and complexity of the required analysis and tests. Some leeway is allowed. If the proposed change is minor, the ACO might waive the need for extensive analysis or testing.

The basic objective of the engineering analysis is to prove that the forces and load limits will remain within the allowable range. For example, in a case where a larger engine is to be installed, the problem isn't restricted to the engine mount. Torque, gyroscopic forces, and slipstream effects, among others, must be considered. The heavier engine might make a beefier landing gear necessary. What about the CG? A shorter engine mount might be required to keep the CG in the proper range; some equipment might be shifted aft to maintain CG. Excessive loads sometimes result in a category change from aerobatic to utility, or utility to normal.

The FAA doesn't care who does the analysis, but most of us lack the right background. Any knowledgeable engineer would be adequate, but the best way is to hire a *designated engineering representative* (DER). Not only can DERs perform the analysis, but they can officially make decisions on behalf of the FAA in regard to whether a particular modification complies with regulations. To qualify as a DER, a candidate must have eight years of aviation engineering experience, plus one year working in aircraft certification with the FAA.

A DER is used to working with the FAA and understands its requirements. In turn, the ACO engineers are familiar with the DER's work. The annual renewal of the DER's appointment, reviews of portions of his or her submittals, and professional pride ensure thorough and precise output. Performance of the work yourself requires an exhaustive review of your submittal by the ACO engineers. This is understandable because they must decide whether the modification complies with the FARs. Since a DER is authorized to make these decisions, approvals are gained far faster.

On the minus side, the DER is a private citizen and charges for his or her services. The going rate is around $50 an hour; the smallest mods will probably take an absolute minimum of at least 10 hours. DERs are appointed for particular specialty areas, such as flutter or structure. Involved modifications might require hiring more than one DER.

Whoever does the analysis will need to know the strength of the aircraft's structure at various points, but the original manufacturer has no obligation to provide this data and has considerable financial and legal incentive not to. The DER might be forced to "reverse engineer" the design, further adding to the expense of the STC.

The fact that another modifier has gained approval for the same modification doesn't help. STC data is proprietary information. The data cannot be used in the approval of another STC unless the owner grants permission. This differs from the field approval where a single precedent can open the flood gates. Each STC application is evaluated solely on the data submitted.

After all, fair is fair. The EAA spent several years and considerable money to gain its autofuel STC. If the FAA had allowed free use of proprietary data, other companies could use the EAA's data to market their own STCs. The EAA would have no chance to recoup its expenses. If this were the case, why would anyone go through the difficulty of gaining approval and marketing an STC?

As mentioned, the purpose of the analysis is to verify compliance with applicable certification requirements. The requirements often demand physical proof as well. The ACO will identify both ground and flight tests that must be performed.

Ground tests can be simple, such as a flow test from a newly installed fuel tank. Or the tests can be quite involved, such as trying to certify a taildragger conversion of a trigeared aircraft (Fig. 10-7). Plan on loading the plane to gross, then hoisting it up and dropping it from various heights to prove the strength of the new landing gear. These tests are dangerous; it is quite possible to overstress the airframe to the point where it is no longer airworthy.

If a modification requires flight testing, the aircraft usually needs temporary certification in the *experimental/research and development* category. The FAA will inspect the aircraft throughout the modification process. Upon completion, the experimental certificate of airworthiness is issued, and flight testing can begin. While these inspections and the certification are free, the ACO's workload might cause delays. If time is a factor, *designated airworthiness representatives* (DARs) can perform the inspections and issue the certificate. But just like DERs, the DARs are private citizens who charge around $50 an hour.

Anyone with a private license can play test pilot. The flight tests verify those details that haven't yet succumbed to the engineer's art, such as engine cooling and handling qualities. The experimental/R&D certificate might require occupant parachutes and the installation of jettisonable doors, especially for spin or flutter testing. When the flight test program is completed, an FAA test pilot will verify the results.

At this point, it's all over but the paperwork. In addition to the engineering analysis and flight test results, the modification must be documented. In the case of one-time STCs, this can consist of detailed sketches (Fig. 10-8) and photographs;

Fig. 10-7. If you want to develop your own taildragger modification, you'll be expected to perform extensive testing. It's easier to just buy a commercial STC package.

however, if trying for a multiple STC, drawings and diagrams sufficient to duplicate the modification must be submitted.

When the FAA signs off, it's done. The aircraft has no restrictions beyond those of its certification category. There is one gleaming star in this whole mess: absolutely no charge by the FAA. It is a small consolation while you're writing out checks to the A&P, IA, DER, and DAR.

Is this trip really necessary?

Sounds arduous, doesn't it? Before you seriously consider modifying your airplane, you'd better make sure you'll be happy with it.

Aircraft design is an exercise in compromise. A long-range airplane should carry lots of fuel; yet the extra weight reduces short-field capability. Airfoils that maximize cruise speed often have vicious stall characteristics. Exotic drag-cheating shapes increase manufacturing expense and the price charged for the completed airplane.

Modifying an aircraft is like poking your finger into a tightly-stacked pile of bricks; one brick might slide back, but another gets shoved out somewhere. The modification process merely adjusts the compromises inherent in the design. It's too easy to focus on the "pluses" and neglect the drawbacks.

For example, examine the gains and losses of installing a larger engine. Cruise speed will increase (although not as much as you might expect), as will the climb rate and the ability to get out of short fields; however, all the airframe-related limitations remain. You might be able to cruise faster, but redline and maneuvering speeds restrict how much of the capability can be legally used. Fuel consumption increases and range gets shorter.

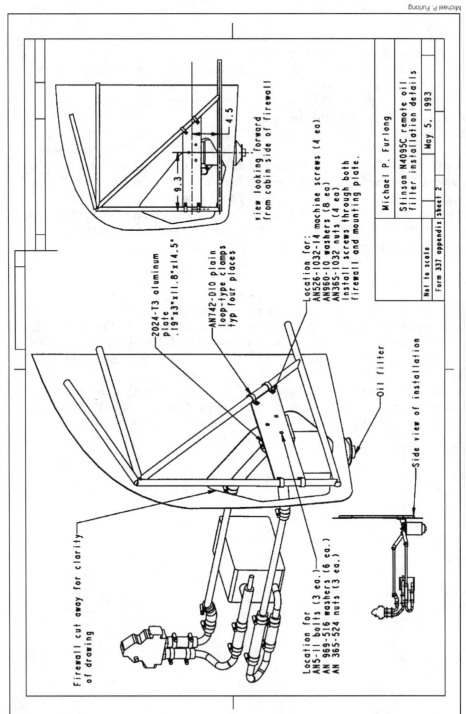

view looking forward
from cabin side of firewall

4.5

9.3

2024-T3 aluminum
plate
.19"x3"x11.8"x14.5"

AN742-D10 plain
loop-type clamps
typ four places

Location for:
AN526-1032-14 machine screws (4 ea).
AN960-10 washers (8 ea)
AN365-1032 nuts (4 ea)
Install screws through both
firewall and mounting plate.

Michael P. Furlong		
Stinson N4095C remote oil filter installation details		
		May 5, 1993
Not to scale		
Form 337 appendix	sheet 2	

Firewall cut away for clarity
of drawing

Oil filter

Side view of installation

Location for
AN5-11 bolts (3 ea.)
AN 969-516 washers (6 ea.)
AN 365-524 nuts (3 ea.)

Fig. 10-8. Each STC application must be accompanied by detailed drawings. They can be done by hand, but must be clear and descriptive. (For illustration purposes only. Not to be used for STC purposes.)

Compensate by adding long-range fuel tanks? This corrects the range problem, but another problem arises. Because the legal gross weight of the aircraft doesn't change, the weight of the heavier engine and extra fuel reduces the useful load. A four-seat airplane becomes a three-seater.

There are also subtle effects that you might not expect. A heavier engine brings the CG forward; a common correction is to move the battery aft to compensate; however, this affects the inertial properties of the aircraft. After entering a spin, it has a greater tendency to keep spinning.

When buying the rights to a multiple STC, get the performance figures beforehand. Do a little mental arithmetic. Assume you're buying a brand-new airplane. Does it still meet your needs? If planning your own STC, talk to an experienced IA or the engineers at the FAA aircraft certification office. They should be able to give some indication of the drawbacks of the proposed modification.

Executing an STC is a lot like making a parachute jump; it's very difficult to get back into the same airplane if you find out you were wrong. Don't let the "pluses" blind you from the "minuses."

Summary

As far as the FAA is concerned, aircraft aren't modified in workshops or hangars. To them, it happens in offices, where the aircraft's type certificate is modified or logbook entries reflecting Form 337 mods are made. The FAA's overall concern is safety, which often has more to do with analysis and test results than with aluminum and steel.

Of course, the person making the modification must incorporate all factors. The paperwork jungle must be plowed; aluminum and steel must be worked. Advance coordination is the key. Talk to the FAA before you start.

11

Owner
maintenance basics

IF YOU'VE EVER DONE ANY AUTOMOTIVE MAINTENANCE, those skills can help achieve a significant drop in airplane ownership costs. In some ways, aircraft are simpler to work on. Changing eight plugs on a '79 V8 Chevy Monza calls for more contortions than doing the same job on a '47 flat-four-Continental Champ. Figure 11-1 shows that maintenance is a lot easier when you can remove a cowling and lay the engine bare in a matter of moments.

Owner maintenance isn't for everybody, though. It *does* take time. Some folks just can't pick up a tool without sticking it through something. I got fumble-fingered recently and ended up dropping a portion of a gas cap into a fuel tank on *someone else's* airplane, no less.

Working on the plane is fun, but sometimes I'd like to just hand my wallet to a mechanic and eliminate the hassle. Unfortunately, my wallet is rarely thick enough to give my fumble-fingers any rest.

This chapter and chapter 12 are for those owners in the same boat. This chapter discusses what the FAA allows the owner to do, what tools are necessary, and some guidelines for aircraft-quality work. Chapter 12 takes a look at various aircraft systems and the owner-maintenance procedures associated with them.

BEFORE YOU BEGIN

Two points must be emphasized. First, there's no shame at being reluctant to perform some owner-maintenance tasks. Working on your own plane can enhance your confidence, but if you're skittish about a particular procedure, you're not likely to be comfortable while flying the plane afterward. By all means, leave it for the A&P.

Second, never imagine that you have to go it alone. Nervous about oil changes? Schedule your first one with the A&P, and have him or her show you how. Similar help is available through type clubs or local owner's groups.

Fig. 11-1. Access to an aircraft engine is much easier than many cars, but it's usually easier when two people handle the cowling.

Even if you don't plan on doing any of your own work, at least skim these two chapters. You'll probably pick up some information that'll make it easier to converse with your mechanic.

Legalities

The authority for an owner to work on his or her aircraft is contained in FAR 43.3 (g): "The holder of a pilot certificate issued under Part 61 may perform preventative maintenance on any aircraft owned or operated by that pilot that is not used under Part 121, 127, 129, or 135."

The term *preventative maintenance* is of interest. The FAA defines it in Appendix A, section (c) of FAR 43. Those sections applicable to normal, utility, and acrobatic aircraft operated under Part 91 are reproduced as Fig. 11-2.

Interpretation, as ever, is the key. Take the first half of the first sentence: "Preventative maintenance is limited to the following work . . .." In other words, if the procedure isn't listed in this section, the owner *cannot* legally perform it.

Simple enough, apparently, but take a gander through the list and point out where it authorizes the owner to perform oil changes. It doesn't. It lets you clean and replace oil strainers, but nowhere does it authorize draining engine oil.

Sure, there's the "lubrication" clause. But my dictionary defines "lubrication" as adding oil, not draining it. Compare how subsections 16 and 17 spell out *exactly* what the owner can do with the position and landing lights, which is a far less crucial system than the oil tank! Fortunately the FAA and the rest of the known world interprets this section to allow owners to change their own oil.

Every time the need for maintenance arises, the owner must make a decision as to whether he or she is legally authorized to perform the procedure. It's not cut

FAR Part 43 Appendix A
(provisions that are applicable to all aircraft categories)

(c) Preventive maintenance. Preventive maintenance is limited to the following work, provided it does not involve complex assembly operations: (1) Removal, installation, and repair of landing gear tires.

(2) Replacing elastic shock absorber cords on landing gear.

(3) Servicing landing gear shock struts by adding oil, air, or both.

(4) Servicing landing gear wheel bearings, such as cleaning and greasing.

(5) Replacing defective safety wiring or cotter keys.

(6) Lubrication not requiring disassembly other than removal of nonstructural items such as cover plates, cowlings, and fairings.

(7) Making simple fabric patches not requiring rib stitching or the removal of structural parts or control surfaces. In the case of balloons, the making of small fabric repairs to envelopes (as defined in, and in accordance with, the balloon manufacturers' instructions) not requiring load tape repair or replacement.

(8) Replenishing hydraulic fluid in the hydraulic reservoir.

(9) Refinishing decorative coating of fuselage, balloon baskets, wings tail group surfaces (excluding balanced control surfaces), fairings, cowlings, landing gear, cabin, or cockpit interior when removal or disassembly of any primary structure or operating system is not required.

(10) Applying preservative or protective material to components where no disassembly of any primary structure or operating system is involved and where such coating is not prohibited or is not contrary to good practices.

(11) Repairing upholstery and decorative furnishings of the cabin, cockpit, or balloon basket interior when the repairing does not require disassembly of any primary structure or operating system or interfere with an operating system or affect the primary structure of the aircraft.

(12) Making small simple repairs to fairings, nonstructural cover plates, cowlings, and small patches and reinforcements not changing the contour so as to interfere with proper air flow.

(13) Replacing side windows where that work does not interfere with the structure or any operating system such as controls, electrical equipment, etc.

(14) Replacing safety belts.

(15) Replacing seats or seat parts with replacement parts approved for the aircraft, not involving disassembly of any primary structure or operating system.

(16) Troubleshooting and repairing broken circuits in landing light wiring circuits.

(17) Replacing bulbs, reflectors, and lenses of position and landing lights.

(18) Replacing wheels and skis where no weight and balance computation is involved.

(19) Replacing any cowling not requiring removal of the propeller or disconnection of flight controls.

(20) Replacing or cleaning spark plugs and setting of spark plug gap clearance.

(21) Replacing any hose connection except hydraulic connections.

(22) Replacing prefabricated fuel lines.

(23) Cleaning or replacing fuel and oil strainers or filter elements.

(24) Replacing and servicing batteries.

(25) Cleaning of balloon burner pilot and main nozzles in accordance with the balloon manufacturer's instructions.

Fig. 11-2. Owner maintenance allowed under FAR Part 43.

(26) Replacement or adjustment of nonstructural standard fasteners incidental to operations.

(27) The interchange of balloon baskets and burners on envelopes when the basket or burner is designated as interchangeable in the balloon type certificate data and the baskets and burners are specifically designed for quick removal and installation.

(28) The installations of anti-misfueling devices to reduce the diameter of fuel tank filler openings provided the specific device has been made a part of the aircraft type certificate data by the aircraft manufacturer, the aircraft manufacturer has provided FAA-approved instructions for installation of the specific device, and installation does not involve the disassembly of the existing tank filler opening.

(29) Removing, checking, and replacing magnetic chip detectors.

(30) The inspection and maintenance tasks prescribed and specifically identified as preventive maintenance in a primary category aircraft type certificate or supplemental type certificate holder's approved special inspection and preventive maintenance program when accomplished on a primary category aircraft provided:

(i) They are performed by the holder of at least a private pilot certificate issued under Part 61 who is the registered owner (including co-owners) of the affected aircraft and who holds a certificate of competency for the affected aircraft (1) issued by a school approved under § 147.21(f) of this chapter; (2) issued by the holder of the production certificate for that primary category aircraft that has a special training program approved under § 21.24 of this subchapter; or (3) issued by another entity that has a course approved by the Administrator; and

(ii) The inspections and maintenance tasks are performed in accordance with instructions contained in the special inspection and preventive maintenance program approved as part of the aircraft's type design or supplemental type design.

Fig. 11-2. Continued.

and dried. An owner is authorized to remove and replace tires, for instance. Nothing in Appendix A allows the owner to remove and install the *tube* inside those tires. Yet when the tires are off, the tubes are trivial.

The second half of the first sentence in Section (c) isn't much help: ". . . provided it does not involve complex assembly operations." What's "complex?" Again, we're at the point where interpretation is up to the aircraft owner.

The FAA doesn't want you making an unwitting mistake that might threaten safety. If you have any questions about the legality of a particular procedure, contact the nearest flight standards district office.

However, an owner is allowed to perform whatever operations on an aircraft that the owner desires, as long as a licensed mechanic is willing to supervise and sign off the work in the aircraft logbooks. As you might expect, this is the sort of thing to arrange *before* taking a wrench to your airplane.

When your mechanic has a feel for your shop skills, you might be able to perform more advanced work. For example, I had a spate of solenoid problems in my old 150. After the third failure, my mechanic let me install the new one at my convenience and just bring the airplane by for an inspection and signoff. There wasn't much to it, just four bolts and a couple of battery cables. He didn't charge me, either.

Restrictions established by the FAA aren't there to generate business for A&Ps. The restrictions make sure that a repaired airplane is *airworthy*, not just reassembled.

The final source for proper procedures and materials is the official service manual. Engine and airframe manuals are available through the manufacturer; for long-gone aircraft, other companies (such as Univair) republish manuals. If you're serious about working on your plane, get the appropriate service manuals.

Primary aircraft

The first aircraft certified under the new *primary aircraft* category started reaching general aviation flightlines in 1994. There are some interesting provisions of the category. Owners of primary aircraft can be authorized to do more than the "standard" preventative maintenance operations. However, you'll have to attend a special school to earn a certificate authorizing you to do so. It's something to check on if you've bought an airplane certified in the primary category.

TOOLS

You'll be needing some tools to assist in your aircraft work:

- Screwdriver set: standard and Phillips. Get an assortment because you'll probably eventually use them all. Power screwdrivers are jim-dandy, especially for removing all the inspection plates at annual time. Be careful, especially while tightening. I once wrenched off a nutplate with an overly enthusiastic electric screwdriver.

- Socket set: U.S. measurement, ⅜" drive with ratchet wrench. A decent set will come with various adapters and extensions. A ½"-drive set might be necessary for some jobs. A ¼"-drive outfit is a bit easier to handle for light work.

- Torque wrench: ⅜" drive. You might be able to work "by guess and by golly" on your Ford, but don't try it on an airplane. There are two kinds of torque wrenches. One has a gauge readout of torque applied. On the second, you set the desired torque and the wrench "clicks" when you reach it. The second kind is easier to use.

- A deep ⅞" socket. That's a *deep* socket, used to remove and replace spark plugs. Aviation plugs are longer than car plugs. To be sure, take an old aircraft engine spark plug to the tool store. (Your mechanic probably has some unairworthy spark plugs lying about.)

- A set of combination open-end and box-end wrenches.

- Safety-wire twisters. These look like largish wire cutters but have a central spiral shaft (Fig. 11-3). You usually have to buy these from an aircraft supplier. The price varies from $25–$75 new. I found a pair at a surplus store for $10.

- Spools of 0.032" and 0.040" stainless steel safety wire.

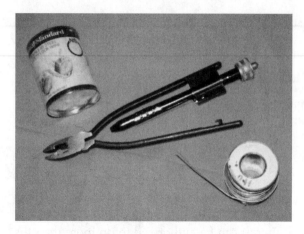

Fig. 11-3. Safety-wire pliers are a required item for most owner maintenance.

- A grease gun and a cartridge of the appropriate grease. Check your service manual.
- Spare screws and other small aviation hardware.
- Pliers, Vise-grips, and other common hand tools.

Most of these can be bought at the neighborhood hardware store. Over the long run, you'll be happier with good-quality tools. In fact, aircraft ownership is a good excuse to buy good-quality tools.

It's most convenient to keep these at the airport in your locker; however, you might be reluctant to leave expensive tools unattended. I leave a small tackle box of cheap screwdrivers and whatnot in my car (Fig. 11-4), and take the good tools when there's real work to be done.

HOLDING IT TOGETHER

There are a lot of similarities between working on your car and working on your airplane. You'll use the same tools, and for the most part the procedures are similar. There is one big difference between car and aircraft work: Aircraft must *not* come apart on their own.

Yes, you don't like your car to come apart on its own, either, but the aviation world is a lot more stringent. Something *must* provide assurance that vibration, heat, or stress will not cause components to separate.

When we bolt something together around the home, we usually just use a nut and a bolt. We tighten the nut thoroughly to prevent it from coming loose. That's not good enough on an airplane. Take a close look at the airplane. Have you ever seen so many self-locking nuts in your life? Also note the cotter pins and safety wire. Aviation hardware is designed to work *and last*.

Bolts

Chapter 10 discussed aviation-quality hardware standards. Most bolts on your aircraft will meet AN specifications. Generally speaking, you'll find AN3-

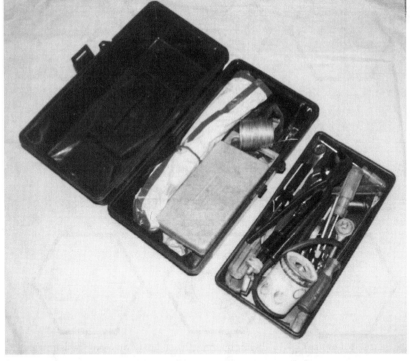

Fig. 11-4. Assemble a portable tool kit for typical light maintenance. It contains a small socket set, a pouch of wrenches, some screwdrivers, the safety-wire pliers, plus wire, tape, and extra washers and nuts.

through AN20-type bolts. AN bolts always have markings on the head to indicate that they are high-strength bolts. A typical marking is a four- or six-pointed star, possibly accompanied by a couple of letters. Typical head markings are shown in Fig. 11-5.

If a bolt doesn't have markings, it isn't an approved bolt and shouldn't be used; however, the markings aren't exclusive to aircraft hardware. The only way to be sure is to buy the bolts through an aviation supplier.

The characteristics of a particular bolt are described by its specification. AN3-14A is a typical specification:

- **AN** means the bolt meets Air Force-Navy standards.
- **3** is the diameter of the bolt in $\frac{1}{16}$ of an inch.
- **14** is a code for the approximate shank length.
- **A** indicates that the thread area isn't drilled for a cotter pin; you won't be able to use it with a castle nut.

Figure 11-6 shows how to decode bolt specs. There are other codes; for instance, a bolt with a head drilled for applying safety wire will have an **H** after the diameter code: AN3H-10A.

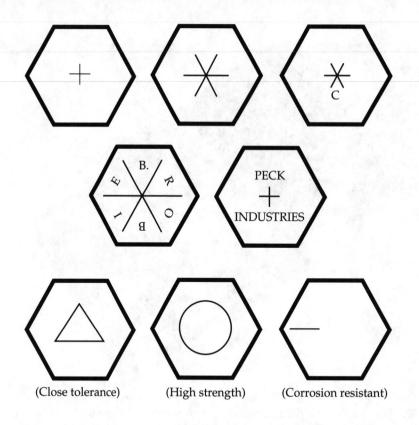

(Close tolerance)   (High strength)   (Corrosion resistant)

Full-Strength Bolts

Substrength Bolts

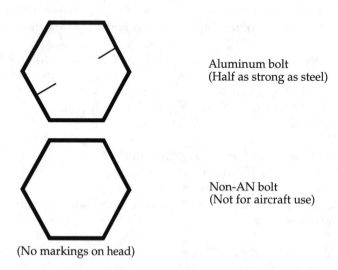

Aluminum bolt
(Half as strong as steel)

Non-AN bolt
(Not for aircraft use)

(No markings on head)

Fig. 11-5. Bolt head markings tell a tale of safety and integrity.

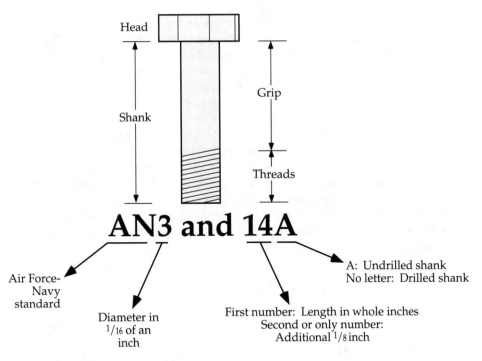

Fig. 11-6. Specifications for AN-type bolts.

The specifications for AN bolts allow some variance for diameter because some bolts might fit tighter than others. When an aircraft designer wants more precision, the designer will specify a *close-tolerance* bolt. These are generally of the AN173 through AN186 series and feature a triangle on the head.

Similarly, sometimes a stronger bolt is required, and there isn't enough room for a bolt with a larger diameter. In this case, the designer might specify NAS1103-1112 series. These bolts feature a cup-shaped depression in the head.

Nuts

The nuts, naturally, are designed to work with the bolts. Most of the "secure" onus is placed on the nut; hence, aviation has a variety of locking or lockable nuts.

The AN365 *elastic stop nut* is an example. It has a semisoft plastic or fiber insert that grips the bolt's threads to keep the nut from turning. Look into the end of the nut to see the colored insert.

This insert can cause some problems. Every time the nut goes on or off, it wears just a tiny bit. Eventually, it loses its self-locking ability. If you can tighten an AN365 stop nut by hand, discard it. Also, the insert is certified to only 250°F. This is adequate for most of the airplane, but they aren't used anywhere near the engine. Instead, you'll typically see the AN363. By using metal fingers, the hole diameter narrows slightly at one end and applies enough friction to prevent self-rotation.

Self-locking nuts aren't the only solution. If the bolt is intended to connect moving parts, the motion can overcome the nut's self-locking ability. *Castle nuts* are the answer. They look like conventional nuts with notches cut out of one side like a castle's ramparts. The notches pass a cotter pin through a hole in the nut. Castle nuts come in two flavors: AN310 for all applications and the low-profile AN320 for applications where the bolt isn't under tension.

Aviation cotter pins in various diameters and lengths are covered by the AN380 and/or the MS24665 specifications. One important point: *Cotter pins cannot be reused.* If you take one off (allowed under FAR 43), replace it with a new one.

You'll want a collection of new cotter pins handy whenever you work on your plane. The collection should include some aviation washers and extra nuts of various types. You won't often have a need for spare bolts, but nuts have an annoying little tendency to bounce away and disappear. It's less irritating to have a batch of spares handy. Pick the dropped ones out of the dust pile the next time you sweep the hangar.

Figure 11-7 shows what some of this hardware looks like and includes some part numbers for common sizes. For $25 or so, you can have a pretty good stock sitting around. I keep mine in old 35mm film containers, but cheap see-through plastic trays with compartments are better.

Using the hardware

There are a few basic rules to remember when bolting things together. When you disassemble a component, make note of the orientation of the bolt head and the number of washers used. Most bolts will have their heads in the same direction; most use only one or two washers under the nut.

If you don't note the configuration and orientation, here are some guidelines: First, bolts generally have their heads up or forward. This provides a "last ditch" defense if the nut somehow comes off; gravity or positive acceleration will keep them in place. There are exceptions to this rule, usually brought about when positioning a bolt the "normal" way interferes with another component. Otherwise, take "head up or forward" as the standard. Take note of any bolts running the "wrong" way before removal. Restore them to their original orientation.

There's no similar rule for bolts running left and right, so memory or written notes will be necessary here as well. Copious notes, sketches, and photos (especially Polaroids) will help, particularly on large projects or full-scale restorations where memories tend to fade. Besides, they're fun to look at later.

When tightening a bolt-and-nut assembly, always hold the bolt stationary and turn the nut. At least one washer should be used under the nut to keep it from grinding the surface of the component. When the assembly is tight, at least one thread of the bolt *must* be visible beyond the nut. If not, replace the bolt with a longer one.

If a bolt is *too* long, the nut might "bottom out" on the threads before it's tight. Try turning the nut without holding the head; if the assembly rotates, the bolt is too long. Remove the nut, and add another washer. Don't use any more than three washers, though.

	To Fit AN3 (3/16") bolt	To Fit AN4 (1/4") bolt	To Fit AN6 (3/8") bolt
Elastic stop nut 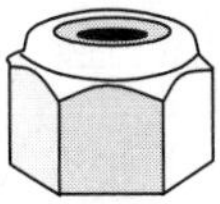	AN365-1032A AN364-1032A*	AN365-428 AN365-428A*	AN365-624A AN364-624A*
Metal stop nut 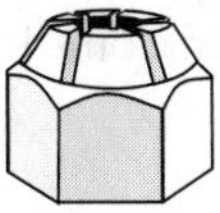	AN363-1032	AN363-428	AN363-624
Castle nut 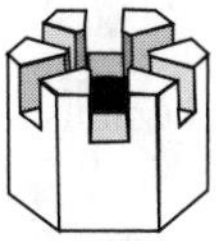	AN310-3 AN320-3*	AN310-4 AN320-4*	AN310-6 AN320-6*
Washer 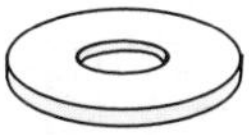	AN960-10 AN960-10L* AN970-3**	AN960-416 AN960-416L* AN970-4**	AN960-616 AN960-616L* AN970-6**
Cotter pin 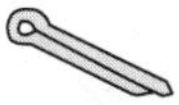	AN380-2-2	AN380-2-2	AN380-3-3

* Thin unit . . . use only when called for

** Extra-wide washer

Fig. 11-7. Typical aviation hardware part numbers.

If things still aren't tight, replace the bolt with a shorter one. *Do not use a die to add threads to the bolt.* AN bolt threads are rolled, not cut; cutting threads causes localized stresses that can cause premature failure. Also, cutting threads on an AN bolt removes the cadmium plating, which gives corrosion a starting point. These basic rules are summarized in Fig. 11-8.

Safety wire

Safety wire secures things that don't have self-locking features. Your oil tank's drain plug will probably have to be safetied in place after an oil change, as will the oil filter or screen and a number of other small items. If it's safety-wired when you start, you'll have to restore it when you're done.

Safety wire comes in several different varieties: soft iron, copper, brass, and stainless steel. Stainless is the most common. Many diameters of stainless wire are available: 0.020", 0.032", 0.041", 0.051", and 0.057". Buy a spool of 0.032" and a spool of 0.041" stainless wire.

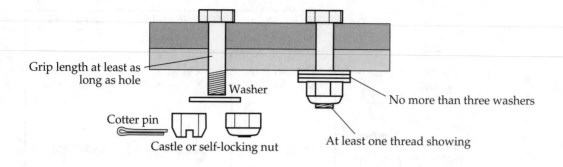

Grip length at least as long as hole

Washer

Cotter pin

Castle or self-locking nut

No more than three washers

At least one thread showing

Bolt heads UP and FORWARD, when possible

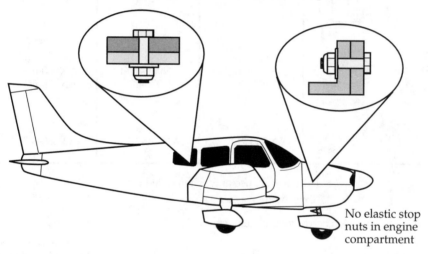

No elastic stop nuts in engine compartment

Fig. 11-8. Basic bolt installation guidelines.

Safety wire material is strong, but not brittle. The wire can take a bit of flexing without breaking. Safety wire isn't designed to carry a load; it is used mostly to keep something from turning, usually a nut. The safety wire will run from the nut to a solid point—sometimes a lug especially for the purpose. Other times it just runs to a convenient projection.

Before cutting off the old wire, make note of what both ends are attached to. You'll want to run the new wire to the same lugs. Discard the old safety wire; never reuse it.

The length of the new piece of wire will have to be more than twice the distance to be spanned, so cut off a generous hunk. If the piece is too short, it invariably doesn't become apparent until after you've spent 10 minutes stringing it through tiny little holes. Cut it long!

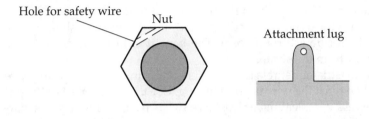

Hole for safety wire Nut

Attachment lug

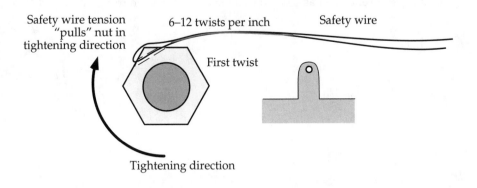

Safety wire tension "pulls" nut in tightening direction

6–12 twists per inch

Safety wire

First twist

Tightening direction

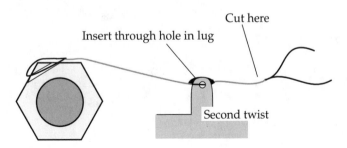

Cut here

Insert through hole in lug

Second twist

Fig. 11-9. Safety wiring sequence.

Safety wire is always installed so that it prevents the safetied component from turning in a loosening direction. This generally means that the wire should curve around in a clockwise direction when viewed from the nut side.

Bend the piece of wire in half, and slide one side of the wire through the hole of the item to be safetied. The ends should be pointing in the "tighten" direction.

Next you'll have to twist the wire from the nut to the approximate distance to the safety point. This way if one leg of the wire breaks, it's still wrapped with a solid piece.

The best way to twist the wire is with a set of *safety-wire pliers*. It's a slick little tool, albeit a bit expensive to purchase. The jaw ends grasp the two sides of the safety wire and the handle locks to hold them. Pulling a center knob spins the pliers and the attached wire. When correctly twisted, the safety wire should have 6–12 turns per inch. Too few turns (loose): The benefit is lost. Too many turns (tight): It might break from the strain.

When the first twist is done, release the pliers and run one piece of the wire through the lug. Then twist the wire beyond the lug. Use the nipper part of the pliers to cut off the excess wire, leaving at least five twists past the lug. Use the narrow end of the pliers to bend the end out of the way. This procedure is summarized in Fig. 11-8.

That should be enough basics to get you started. Now, on to the *real work*!

12

Common owner
maintenance procedures

IF YOU ARE READY AND WILLING, this chapter contains information to help you get started on the most common owner maintenance operations. You are strongly encouraged to find an experienced person to check your work the first couple of times. If your mechanic is willing and convenient, great; otherwise check around for another owner doing a similar procedure.

A good source of unofficial assistance is your local chapter of the Experimental Aircraft Association. Homebuilders get quite experienced in maintaining their own planes, especially because the original builder of an aircraft can perform all the functions of an A&P.

The subsections summarize various systems on aircraft and discuss the preventative maintenance functions that the owner can perform. One caution: The simplified diagrams included aren't based on any particular design. The diagrams are meant to help you understand how the system works. Consult service manuals for the specific diagrams appropriate to your plane.

ENGINE OIL SYSTEM

From the owner's point of view, the oil system and the fuel system are two of the most important elements of the airplane. Why? Because a mistake during preventative maintenance can lead to sudden engine failure. It behooves an owner to understand the *whys*, and not just the *whats*.

Oil system basics

A schematic representation of an aircraft oil system is shown as Fig. 12-1. This isn't representative of any particular brand of engine; rather, it's a collection of typical oil system elements that I'll use to explain system operation.

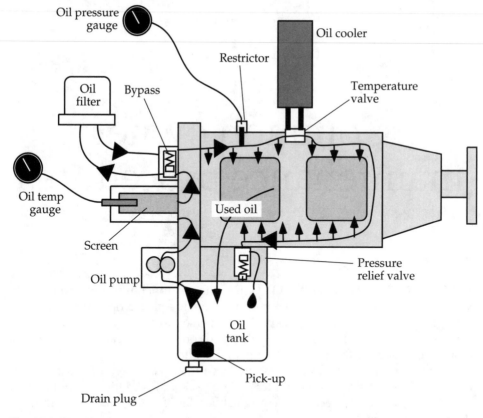

Fig. 12-1. Simplified oil system diagram. On some engine types, the pressure relief valve is located in parallel with the oil pump.

It all starts near the bottom of the oil tank. Oil enters the system through a pickup tube. This tube usually has a rather coarse strainer at the end to filter out large items that could plug the oil passages.

Suction to pick up the oil is provided by an engine-driven oil pump. (The back of the engine contains gearing to turn a number of items, like the oil pump, magnetos, vacuum pump, and generator. It's called the *accessory case*.)

Oil leaves the oil pump and travels through the oil screen. This is a rather coarse metal-mesh filter. *Light Plane Maintenance* magazine calls it the "Elephant Catcher." It gets the large debris out of the oil, but none of the fine grit. The screen is a permanent item, not a throw-away element.

Obviously, engines stay healthier when the finer grit is filtered from the oil. Some engines include a spin-on oil filter similar to one on a car. The spin-on filter can attach directly to the engine or be connected by hoses as a "remote-mount" filter. Remote-mount units are typically added on after engine manufacture.

A filter is much more effective than a screen. On most engines, the oil change interval *doubles* when a filter is installed. If time pressures make you have the

A&P do the oil changes, you can quickly recoup the cost of the STC for a remote-mount filter.

One problem can occur, though. Because the spin-on filter has a finer mesh, it is more prone to clogging than the Elephant Catcher. To prevent a clog from stopping oil flow, aircraft filter attachments include a *bypass valve* that switches the filter out of the oil flow if it becomes too restrictive.

A similar valve is placed at the base of the oil cooler, if one is installed. This valve works slightly differently; if the engine oil is cool, it shuts off oil flow to the cooler. As the oil warms, the temperature valve shunts more and more oil through the cooler. An A&P can adjust the valve to set the desired temperature.

Yet another kind of valve is installed either near the oil pump or well downstream. The *pressure relief valve* is used to control the overall oil pressure. Some systems block oil flow until it reaches the desired pressure. Other systems shunt oil back to the input of the oil pump when the pressure attains the set point. If your oil pressure tends to run too high or too low, it might merely be a misadjusted pressure valve. Ask your A&P to check.

The FAA requires an oil pressure gauge in the cockpit. It gets its information from a tap in the oil flow. Lycomings generally put the *pickoff* near the oil pump; Continentals tend to put it well downstream.

The difference can be seen during startup. A Lycoming will show oil pressure within seconds. A delay of up to 30 seconds isn't unusual in a Continental. It doesn't mean that the Lycoming has a better oil pump; if the Lycoming's pickoff were placed at the same relative location as the Continental, it would probably show the same behavior.

Two types of oil pressure gauges are found on airplanes. One places a transducer on the engine; the transducer sends the pressure to the cockpit gauge electrically. The second more-common type is a mechanical gauge connected to the engine oil system via a tube.

If this tube were to break, however, oil would start pumping out of the engine. To minimize this hazard, a small-diameter tube is used or a restrictor is installed at the pickoff. The small size doesn't affect the pressure reading, but would reduce the rate of oil loss if the tube broke.

It's important to understand the oil pressure system in your aircraft. If the tube breaks, the gauge will drop to zero but the engine still has oil; the break won't adversely affect oil pressure for a while.

A friend's airplane lost oil pressure while flying over a large bay, and hot oil started spraying his legs. An A&P, he correctly assumed the gauge line had broken. He knew he had some time until the engine actually seized, and he was able to reach an airport on a nearby island.

That's not to say that an indicated loss of oil pressure isn't an emergency. If the problem is a broken oil-pressure-gauge tube, the engine is still going to eventually run out of oil. If there aren't any other signs of distress (rise in oil temperature, strange noises, visible oil on the outside of the aircraft, etc.), consider pulling the throttle to idle rather than shutting the engine down; it might save the day if you inadvertently start undershooting the forced-landing target. Don't take chances with the souls on board, though.

Changing oil

With some small exceptions, changing aircraft oil is similar to changing oil on your car. To begin, warm up the engine. Do a couple of touch-and-goes if the weather's nice; otherwise idle it for 10 minutes or so—away from other people's airplanes, thank you. Not only does warm oil flow better, but running the engine stirs up the unfiltered debris so that it'll flow out with the old oil.

When you're back at the hangar, remove the cowling. Position a bunch of old newspapers under the engine and have a roll of paper towels standing by because you're likely to need them.

The oil tank contains a drain and drain plug at the bottom. You'll see two types of plugs. One is an automotive-style short bolt that must be unscrewed to drain the oil. The second is a short valve fitting that doesn't require tools (Fig. 12-2). It consists of a small metal assembly with a short metal tube jutting downward.

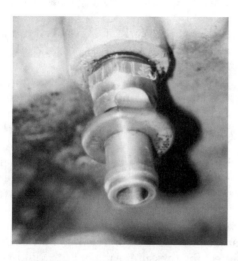

Fig. 12-2. Oil quick-drain valves make oil changes easier and less messy.

If you've got the plug-style, take your side cutters and remove the safety wire. Position a pan or pail underneath the tank. If you can rig a funnel with a short piece of hose, do so because oil tends to splash around. Take your wrench and loosen the plug; when it is loose, remove it by hand. You should be able to sense when it's almost free. Make sure you hang onto it; fishing for a dropped drain plug in a pan of hot, dirty oil isn't much fun.

If you've got the valve-style drain, rejoice. Put the pan under the engine, and slip a piece of garden hose over the metal tube. The hose should be long enough to reach the pan. When the valve is opened, the oil flows down the tube to the pan, no muss, no bother. How do you open the valve? Generally, the collar around the fitting pops upward. On my old 150, it had to be turned 90° first. Ask your mechanic.

Leave the oil to drain and drip while you take care of the screen and/or filter. Most engines have the screen on the back of the accessory case. A screen can screw in or sit behind a bolted cover. Make sure you have a replacement gasket, if required.

Position a pail or some wadded-up paper towels under the screen location. Cut the safety wire and unscrew/unbolt the screen. You might need a pretty big wrench. Often the oil-temperature sensing bulb is screwed into the end of the screen housing; the sensing bulb has to be removed before the screen. On the small Continentals, hold the screen stationary with the big wrench while turning the bulb fitting.

When the screen is off, check the contents. Flush it with a little solvent (a tiny squirt of gas will do), and look for any metal pieces. Reinstall it and safety-wire.

If your engine has a filter, you might not have to remove a screen. Check with your mechanic. Finding the screen the first time that you want to change the oil might be a bit challenging; however, locating the filter is simplicity itself: big, round, and typically white.

Before proceeding, take careful note on how the safety wire runs, then clip it away and unscrew the filter. If it's reluctant to move, put a socket on the end hex fitting or use an automotive strap-type filter wrench.

Some engines don't like sitting around too long without an oil filter; they lose the prime in the oil system. So it's best to put on the new filter fairly quickly. First, clean off the filter mount on the engine. Make sure the old gasket is completely gone.

Hold the new filter open-end-down and slap it a couple of times to knock out any junk that might have worked its way inside. Lubricate the end gasket with a little oil or Dow-Corning DC-4 lubricant (Fig. 12-3). If the filter remains upright so that oil does not spill out, pour new oil into the filter immediately prior to installation. If the filter mounts to one side, you might consider pouring in just enough oil to soak the elements of the filter because any excess oil will spill out when the filter is installed.

Fig. 12-3. Before installing the new oil filter, lubricate the end gasket.

Position the filter in place and screw it on until the gasket contacts the mount (Fig. 12-4). The amount of torque required should be specified on the filter itself; most require about 18–20 foot-pounds. For those without a torque wrench or the right-sized socket, *Light Plane Maintenance* magazine recommends turning the filter for 1 and ¾ turns after the gasket touches the mount. Safety-wire the filter, and that job's done.

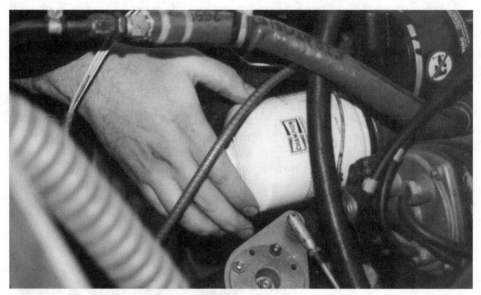

Fig. 12-4. Most oil filters spin-on just like car filters, except the airplane's filter must be safety-wired in place. The filter should be torqued to 18–20 foot-pounds.

Now turn your attention to the old filter. It really should be cut open and examined. Cutters, though, are pretty expensive, about $100 for the Champion tool. Get a couple of fellow aircraft owners to chip in some money and buy one. Some owners just have the mechanic open the filter at annual time.

There is an AD out on some engines that *require* cutting open the filter every 50 hours. If you've got one of those engines, the A&P will have to do the honors and make the appropriate notations in the logbook.

The second-to-last step in your oil change is to reinstall the sump's drain plug and safety-wire it in place (or close the valve, if so equipped).

Finally, add the right amount of oil to the tank. Not every aircraft engine *wants* to be full of oil. Many manufacturers recommend that the tank be filled to the "Full" line only when the aircraft is about to leave on a cross-country. Engines include a *crankcase breather* line to vent any pressure in the crankcase. Some engines will happily spew oil through this tube as the airplane climbs or descends; hence, the manuals call for a full oil tank only if the plane's going to stay at a constant altitude for a while.

Sure, you can fill them to the "Full" line, but eventually you'll probably be scrubbing oil off the belly. Follow the instructions in your manual.

One other caution: Oil comes in little plastic jugs rather than cans, which is a nice setup because we no longer need to chase down a metal spout and ram it through the top of the oil can. One little problem: When you open the plastic jug, a little plastic ring separates from the cap. The ring sometimes remains on the top of the jug, even when it's tilted to pour the oil out. But sometimes it slides down the jug, into the oil filter, and into your oil tank. Oh, oh. If the ring stays with the jug, *take the ring off* before pouring the oil. (Most of the oil jugs that I've seen lately have a split-ring that stays with the cap.)

When the oil tank is full, replace the filler cap. Check that the plug is in place, and the plug, screen, and filter are safety-wired. Then crank up the engine.

Watch the oil pressure. It might be a little slower to come up, but not much. If it isn't off the peg by the manufacturer-stated time, shut down. Otherwise, run the engine for a few minutes, then shut down and look for oil leaks. If it's clean, button up the cowling and go for a short test flight.

FUEL SYSTEM

The fuel system is as important as the oil system, even more so, perhaps. The FAA allows very little owner maintenance on it. You can "clean or replace the fuel . . . strainers and filter elements," and that's about it.

Like everything else, this is subject to interpretation. After all, most aircraft carburetors include a fuel screen deep inside, but I bet the FAA would just as soon we didn't start disassembling our carbs.

Fuel system basics

A schematic of a low-wing-aircraft fuel system is shown in Fig. 12-5. Gas flows from the tanks, through the selector valve and gascolator, and into the carburetor, all the time hurried along by the main (engine-driven) and auxiliary (electric) fuel pumps.

High-wingers eliminate the fuel pumps and use gravity to move the fuel. The distance from the bottom of the tank to the carburetor is called the *fuel head*. There must be sufficient head to allow the system to provide 150 percent of the engine's full-throttle fuel burn; otherwise a fuel pump must be installed. Even a high-winged aircraft might have a pump if it doesn't generate enough fuel head or if a fuel-injected engine is installed. There are other permutations. Some planes pump gas to a *header tank* that is in front of the cabin, and the gas "gravity-flows" to the carburetor from that tank.

One oft-asked question: Why don't low-winged aircraft's fuel selectors have a BOTH setting? Two glasses, a couple of straws, and a bottle of your favorite libation will provide the answer.

The glasses represent the fuel tanks, the straws are the fuel lines, and you get to be the low-winged airplane's fuel pump. Fill one glass, leave the other empty,

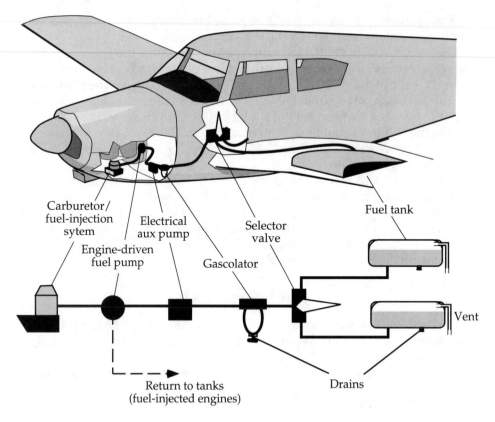

Fig. 12-5. Simplified fuel system schematic.

and place the glasses side by side. Put a straw in each glass. Put your mouth over the top end of both straws, and suck on both at the same time.

You get a mouthful of air. Pulling a liquid uphill is a lot tougher than sucking air. The pump sucks air instead of liquid. Use your tongue as a "selector valve" while considering how to simulate a high-winged plane. You'd take both glasses— the full *and* the empty one—and turn them upside down over your head. "Fuel" will flow, all right.

Fuel system maintenance

The main "strainer and filter" that you're allowed to work on is the gascolator, usually located low on the firewall. It's not something you'll have to mess with very often. For most owners, cleaning it at every annual will be sufficient; however, if you discover sediment or foreign matter in the fuel, open and clean the gascolator.

Begin the process by turning the fuel valve off. That "150 percent of full-power fuel burn" is no fun at all if it's cascading onto your hangar floor. Next, position a bucket under the gascolator to catch the gas that'll trickle out of the bowl and the lines. Open the drain on the gascolator to start the flow.

The gascolator bowl can be removed without taking the unit from the aircraft or disconnecting the fuel lines. There's usually a small nut or thumbwheel on the bottom on a metal *bail* (thick wire) slung underneath the bowl. Cut the safety wire on the thumbwheel and start turning. The bowl will gradually slide down. Sometimes it comes low enough to slide out without disconnecting the bail; otherwise, just pull the tops of the bail out from the upper housing. Often the gascolator doesn't have a bowl *per se*; top and bottom housings are separated by a 2-inch length of aluminum or glass tube.

The first time I opened the gascolator, I was sure I'd been gypped. Nothing was inside! I was expecting a big element like a car fuel filter. A careful examination found a skinny stainless-steel screen underneath the housing (Fig. 12-6). Other types do have a larger canisterlike filter screen.

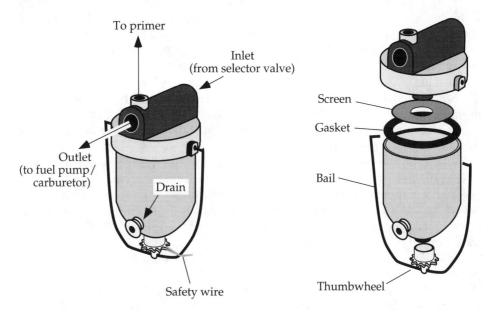

Fig. 12-6. This gascolator is similar to the unit that is installed in author's club airplane. Many aircraft use a canister filter instead of the small stainless-steel screen.

Remove and clean whichever type you have. Wipe out the inside of the bowl and the top housing. Reassemble the unit the same way it came apart, making sure the neoprene or cork gasket atop the bowl is properly positioned inside the upper housing. Tighten the thumbwheel and safety-wire.

Turn on the fuel and check for leaks. Run the engine for a minute or two and check again. A leaky gascolator is dangerous. Not only can the loose fuel catch fire, but air can be introduced into the fuel line to the carburetor. Engines don't run well on air.

In addition to the carburetor screen mentioned earlier, there are other screens in the fuel system. Each tank, for instance, includes a *finger strainer* at the outlet port. You probably won't get involved with these unless you're restoring the aircraft.

As a side note, the mesh on the fuel system filters gets progressively finer from the tank to the carb. The finger strainer is relatively coarse, the gascolator is finer, and the final screen within the carb is fine indeed.

BRAKES AND WHEELS

As mentioned earlier, owner maintenance on the wheels and brakes is another interesting FAR 43 interpretation issue. The FAR doesn't include any brake servicing under preventative maintenance. Yet, if your plane has disc brakes (like most aircraft), you'll have to remove the brakes when changing a tire. When the brake assemblies are off the landing gear, it's often a very simple job to replace the brake shoes, yet the regulation doesn't allow you to.

Why use such a system, anyway? You don't have to remove the brakes on a car if the tire goes flat! It all boils down to saving weight. A disc brake works by clamping two pads on either side of a metal disc attached to the spinning wheel. The disc on a car is attached to a hub that bolts to the axle and incorporates the necessary bearings. The wheel then bolts to this hub. A car's flat tire won't strand you on outskirts of nowhere; you can quickly unbolt the wheel and slap on a spare.

Rather than a separate hub and wheel, aircraft save weight by combining them in one light unit. It's not a major concern safetywise. A flat tire on your plane doesn't leave you stranded in a cow-skull-studded desert. In most cases, the flat happens at a nice, safe airport.

Like cars, not all aircraft use disc brakes. Various drum systems are used. These can often be removed without disturbing the brake system.

You're allowed to replenish brake fluid. Sometimes the reservoirs are easily located (Fig. 12-7), and sometimes they're buried under the instrument panel as part of the brake master cylinders. One no-mess system for adding brake fluid is by using a pump-type oil can.

One warning: Use *only* aviation brake fluid. The automotive stuff is *not* compatible with aircraft systems.

Before we get started, let's get our nomenclature straight. When I say "tire," I mean the stiff black rubbery thing that's in contact with the asphalt. "Wheel," technically, is the metallic thingamabob that the tire is wrapped around. I'll use the term "rim" for the edge of the wheel closest to the outside, where rubber meets metal. The "tube" is the flexible rubber inner tube. The "valve" is the part where you add air; the "stem" is an extrusion (from the tube) that holds the valve.

Now, let's go through wheel removal and disassembly.

Jacking

The process begins with figuring out how to jack up the airplane. Park the airplane on a flat spot away from grease spots or loose gravel. Each wheel that is not going to be jacked should be chocked fore and aft.

Next, remove any safety wire or cotter pins, ensuring that you have replacements handy. Sometimes it takes a lot of jerking and pulling to get the cotter pins out. For safety's sake, do it before the airplane is up on jacks.

Fig. 12-7. The brake fluid reservoir of this Piper Arrow is easy to get to. This is not the case with many aircraft.

A variety of formal and informal methods are used to jack the airplane. Your service manual will show the approved methods (Fig. 12-8). Usually, there'll be *hard points* where a standard hydraulic bottle or floor jack can be placed. Sometimes the tiedown ring is identified as the hard point; the ring is unscrewed and replaced with a special jack fitting (Fig. 12-9).

Other planes use special adapters. Bonanzas were delivered with fittings that slid into the back side of the gear leg to allow jacking. If a tall wing jack isn't available for your Cessna, various adapters clamp to the gear legs. If you're changing a nosewheel, a tightened tiedown rope at the tail might be just the ticket.

Before you hoist away, consider *what's going to happen* if the jack slips or suddenly goes flat? When the wheel comes off, slip a block of wood, stack of 2 × 4s, or old milk crate under the bare axle, just in case (Fig. 12-10).

Hoist away when ready. Watch the base of the jack; make sure it doesn't start sliding. If you're jacking spring-steel (Cessna) or bungee-style (early Piper and Aeronca) gear legs, be wary. The leg will tend to splay outward a bit and tilt the jack. Use a rubber mallet or piece of wood to tap the base of the jack back into place.

One last point: Be careful when jacking taildraggers. If you place the jack behind the wheel, the plane might tip forward on its nose.

Removing the wheel

Wheel removal is fairly straightforward. The primary thing is to remember the order it came apart. Draw a sketch, or lay out the parts in sequence.

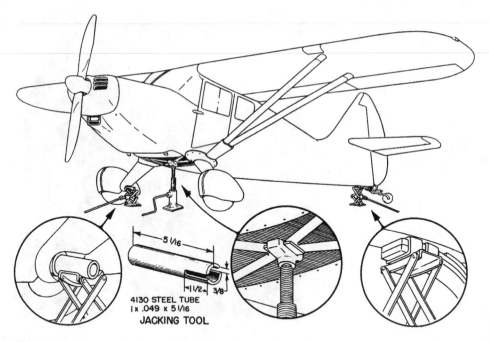

Fig. 12-8. Jacking methods for the Stinson 108. Univair Aircraft Corporation

5 1/16

4130 STEEL TUBE
1 x .049 x 5 1/16
JACKING TOOL

1 1/2   3/8

Fig. 12-9. This Mooney is being jacked up using owner-built stands and a standard bottle jack. Tiedown rings under the wing are removed and replaced with tapered pins that fit a recess in the jack shaft.

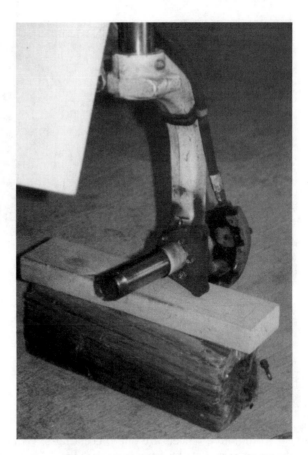

Fig. 12-10. Whenever the wheels are off, place something under the axle to prevent damage to the plane in case the jacks slip.

The hardest part is getting the brake calipers out of the way. The caliper is a U-shaped device wrapped around the brake disc. It's held in place by (typically) two bolts. Some planes have rubber brake lines; on these planes, the whole caliper can be unbolted and removed (Fig. 12-11). It shouldn't be necessary to remove the brake hose, but use a bit of tape or wire to hold the caliper so that it's not just hanging on the hose.

With planes having steel lines, you'll split the caliper in place. Whichever way, there might be some metal shims between the two halves of the calipers; remember which goes where.

Wheels are held in place by a variety of methods. Those installed on forked legs usually have a long bolt through bushings in the fork. Otherwise, the end of the axle is probably threaded for a large nut. These are shown in Fig. 12-12.

When ready, remove the axle bolt, nuts, and the like, taking careful note of their order. At that point, the wheel should be free. Slide and wiggle it off the axle.

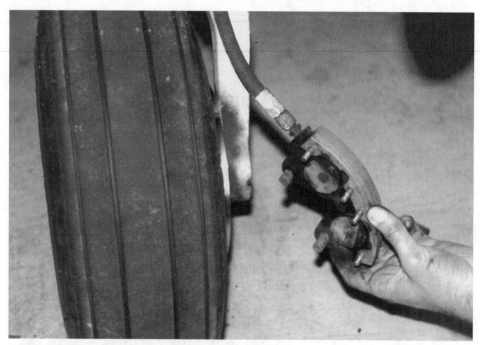

Fig. 12-11. When the through-bolts are removed, the caliper can be lifted away from the wheel. Note the light-colored brake pads.

Removing the tire

Automotive wheels are one-piece affairs. While simple and cheap, they complicate the removal of the tire from the wheel. If you've watched a tire shop at work, you've seen them use a "bead breaker." They ram this angled piece of steel between the inner part of the tire (the "bead") and the wheel and use a motor-driven system to drag it around the wheel. It pops the bead to the outside of the wheel.

Aviation? We just take the wheel apart. Start by marking the two sides of the wheel so you can reassemble them at the same orientation. This'll make sure it's still in balance when reassembled. Mark the brake disc, as well. The tire will have a stripe or dot painted on it to indicate where it is lightest. The stripe or dot is to be aligned with the valve stem when the wheel is reassembled.

Next, let the air out of the tire. The easiest way to do this is to remove the valve core. The simple little tool to do this is available at any auto, bicycle, or hardware store. Pick up a pack of replacement cores, too. Often a slow leak is caused by a faulty core; you can replace the core and pump up the tire without removing it from the airplane.

When the air is out of the tire, you'll need to separate the bead away from the wheel's rim. All you have to do is get the tire loose. You don't have to flip the edge of the tire over the rim.

To break the beads on my club's homebuilt, I use four simple, commonly available tools: two knees and two palms. I set the wheel flat on the floor and kneel on

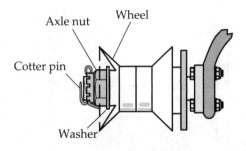

Axle nut   Wheel

Cotter pin

Washer

Axle/axle nut size exaggerated
for clarity

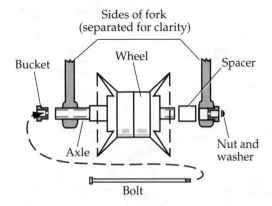

Sides of fork
(separated for clarity)

Bucket   Wheel   Spacer

Axle

Nut and
washer

Bolt

Fig. 12-12. Typical mainwheel (top) and nosewheel.

the tire, evenly spacing my knees and palms around the circumference. Then I kind of jiggle up and down until the tire pops away from the rim. Then I flip the tire over to separate the other side.

However, I'm a tad heavier than many of you. If worst comes to worst, try standing on the tire with your feet on either side. *Carefully* hop until the bead breaks away.

You can avoid this hassle with a bead breaker. The aviation variety is a lot more civilized and cheaper than you'll find at the local gas station. What you *don't* want to do is stick anything sharp (like a screwdriver) between the rim and the tire because you would probably gouge the wheel and quite possibly puncture the tube. Tubes can be patched, but wheel gouges can lead to cracking.

From this point on, it's easy. Most aviation wheels consist of two halves bolted together (Fig. 12-13). While similar, one half will have a hole for the tube stem. Note whether the bolt heads or the nuts go on the stem side, then unbolt. The tire will slide clear of the wheel halves. Be careful not to snag the stem. If you've got disc brakes, the disc comes off with the bolts.

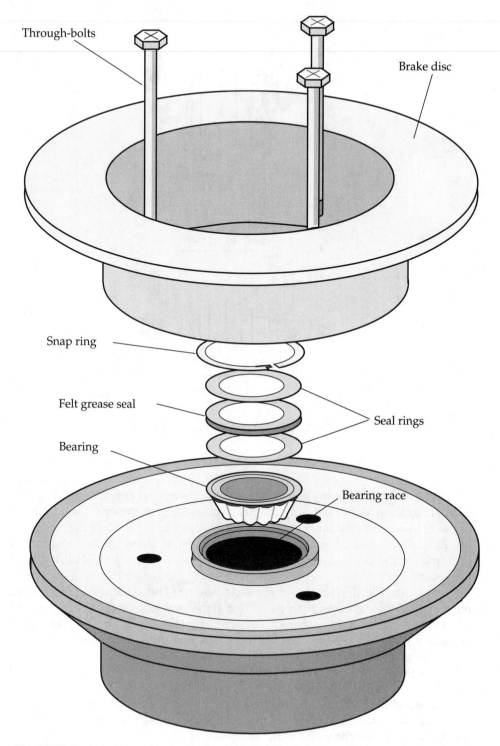

Fig. 12-13. Exploded view of typical general aviation wheel. Larger wheels use six through-bolts.

Other types of wheels come apart similarly. Cub wheels, for instance, have a snap ring on one side; when a retaining pin and the snap ring is removed, the end rim slides right off.

When the tire's off the wheel, reach your hand into the opening and carefully slide the tube out. Do this slowly; even with the valve core out, the tube is sometimes reluctant to collapse.

Wheel bearings

Anytime a wheel's disassembled, you might as well inspect and regrease the wheel bearings. Generally the service manual will call for regreasing at every annual. To some extent, it depends on how much your plane taxis.

One part of bearing work appeals to the kid in all of us: You get to play with lots of messy axle grease. One warning: Keep the grease away from the tire and tube. Petroleum products hasten rubber deterioration.

Generally, a snap ring holds the bearing and spacers in place. Remove the snap ring with needlenose pliers or the tip of a screwdriver. Take off the spacers and any felt grease retainers, taking care to remember the sequence. I just stack them in the proper sequence and reinstall all at one time.

Cleveland wheels use a slightly different system. The rings are held on with three small screws. Removing them frees the assembly.

The bearing should slide out of its *race* (its groove). Wipe off the surface grease and take a close look. Your inspection shouldn't reveal any flat or loose rollers or any corrosion. Any cracks or deformations mean that it's time to get a new bearing. Ditto if the bearing grates when turned or the rollers seem loose. If you need a new bearing, try to read the make and part number off the old one. Call a local bearing distributor rather than an aviation parts shop.

Now comes the fun, greasy part: repacking the bearing. Some references call for cleaning out the old grease with solvent. Not a bad idea, but one that's fraught with danger. You must ensure that the bearing is *completely dry* before packing new grease into it. Any drops of remaining solvent will keep the new lube from sticking, which would not be good for the bearing. I generally just clean off the surface and pack new grease in until the old stuff is forced out.

Open your can of grease (check your service manual, but it'll probably be the same stuff sold at the auto-parts store), and slap a large dollop into the palm of one hand. Cup your palm around the mound. Hold the bearing in your other hand, and press the tapered (roller) side into the grease (Fig. 12-14).

The new grease is forced in between the rollers. You should see a little bit of darker old stuff ooze from between the inner and outer race. Repeat until new grease oozes out. Then rotate the bearing a little bit to bring another set of rollers around, and shove it into the grease again. Work the bearing around, adding more grease to your palm when necessary.

It is fun work, but I will be honest; there are tools that pack the bearings with an ordinary grease gun.

When done, clean the grease out of the race and slip the bearing into position. Wipe your hands clean on a paper towel, then restore the seals and spacers to their

Fig. 12-14. Greasing wheel bearings is easy but messy.

position. Flip the snap ring into place, then repeat the process on the bearing in the other wheel half.

Wheel reassembly

Before putting things back together, it's a good time for an inspection. Wipe everything clean. There shouldn't be any grease visible; the bearings depend on the grease *inside*, not outside.

If you disassembled the wheel due to a flat tire, the problem undoubtedly lies in the tube. Happily, there's no magic technology involved; any old tube repair kit will suffice. Have a new tube standing by in case the old one is unfixable.

Start the reassembly by slipping the tube back into the tire. It's a tight fit, but don't spray any kind of lubricant on it! Many so-called "silicon" sprays include oil, too. Instead, use baby powder. Dash it liberally over the tube and inside the tire. Bunch up the tube and slip it into the tire. Reach in from the other side to pull—gently!—and guide it into position. If it starts to jam, dump some more baby powder on it.

Check the tire indicator (white stripe or red dot) and slide the tube around until the indicator is properly aligned with the tube's inflation valve. On my tires, the dot must be aligned with the valve. Some tubes also have stripes painted on them; use this stripe as a reference instead of the valve.

When the tube is in place, pump it up just a little bit, then let the air out again. This gets the last little creases and whatnot out of the rubber.

Slide the wheel half with the valve cutout into the tire. Place that side down, then slide the other half into place. Look carefully at the mating surfaces of the wheel halves. They should be meeting exactly. If there is a gap, perhaps the tube is pinched. You should be able to hear the metal wheel halves grind against each other.

Rotate the stemless wheel half until your alignment marks coincide, then bolt the wheel together. Don't forget the brake disc. When the bolts are all lightly installed, gradually tighten each one by following a criss-cross-pattern just like you should tighten the lug nuts on a car. Torque values should be stamped on the wheel for easy reference.

When everything is tight, inflate the tire to rated pressure, and let the air out. This again allows the tube to shift around and get comfortable, which is why we put baby powder inside the tire as well as on the tube. Then install the valve core and reinflate.

Reinstalling the wheel

With the wheel reassembled and inflated, it's time to put it back on the airplane. Wipe a small amount of lightweight oil over the axle to lubricate it and slow corrosion. Slide the wheel into place and reassemble the axle assembly.

If the manual shows an appropriate torque for the axle nut, go ahead and apply it. Otherwise, tighten the nut while spinning the wheel. You want it tight enough so there's *just a little drag*. The wheel should coast a little when you let it go, but should drag down to a halt rather than spin on forever. Most people tend to install the wheel too loose, not too tight.

Install a cotter pin (rotating the nut backward slightly, if necessary, to get the castle nut lined up with the hole in the axle), and reinstall the brake calipers.

The next time you come out to the airport, have your bicycle pump or compressor handy. Newly assembled wheels will tend to lose a little pressure as the tube stretches into place.

Rotating the tires

Because they're used only for a few minutes per flight, plane tires don't wear out as fast as car tires. Our club homebuilt goes three or four years before new tires are necessary.

Some effort is necessary to help tires last as long as possible. On many aircraft, unless the suspension is perfectly set up and matched to your typical flight load, tires will develop unusual wear.

Figure 12-15 shows a typical Cessna spring-steel landing gear. To absorb the shocks of landing, the gear legs are designed to move through a considerable arc. The legs hang down somewhat in flight, but during a hard landing, they'll splay out considerably.

Tires last the longest when they run perfectly vertical to the ground. But this condition can only be achieved at a particular load level, usually at gross. If you fly your 172 solo most of the time, its landing gear will tend to be a little bow-legged. The main gear tires' tops will be farther apart than the bottoms, a condition tire people call *positive camber*. It's harmless, from a control point of view.

Unfortunately, as Fig. 12-15 illustrates, camber moves the wear areas of the tires outward. This results in one side of the tire wearing more than its other side. The solution is obvious. Periodically, you've got to rotate the tires to let the other side wear

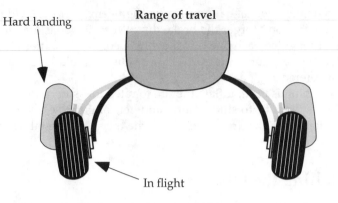

Range of travel

Hard landing

In flight

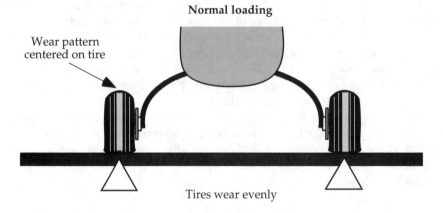

Normal loading

Wear pattern
centered on tire

Tires wear evenly

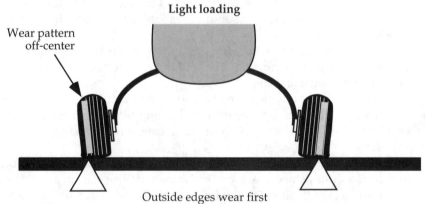

Light loading

Wear pattern
off-center

Outside edges wear first

Fig. 12-15. Light loading or misadjustment of the gear can cause uneven wear of maingear tires.

for a while. Just changing wheels from left to right won't do it. The tire must be physically removed and flipped over, as Fig. 12-16 shows. Remove and separate the wheels, and reassemble with the opposite side of the tire on the stem side.

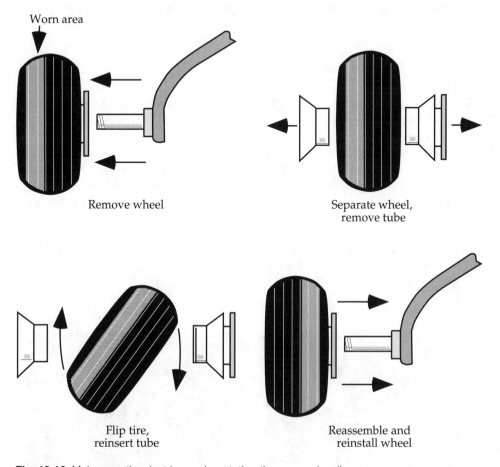

Worn area

Remove wheel

Separate wheel, remove tube

Flip tire, reinsert tube

Reassemble and reinstall wheel

Fig. 12-16. Make your tires last longer by rotating them occasionally.

About that brake disc

Most disc-brake equipped airplanes, like cars, come with steel brake discs; however, unlike cars, airplanes don't use their brakes that often. The steel discs will tend to rust between uses.

If your plane flies regularly, or the local climate is dry, this won't be much of a problem. Otherwise, though, the surface of the discs will become roughened with corrosion. This isn't so much of a safety problem as a pad-wear one. The rough surface will tend to eat the brake pads, which will have to be replaced much more often. If corrosion is an issue, consider replacing the steel discs with chrome or stainless-steel units. Your mechanic can give you more information.

IGNITION SYSTEM

The one allowed owner-maintenance function on the ignition system is the removal, cleaning, gapping, and replacement of spark plugs. An understanding of the overall ignition system will foster a better understanding of spark plug specifics.

The magneto

I have a confession to make. Several of my friends are installing automotive engines in their homebuilt aircraft. One of their reasons is to eliminate that "hard-starting, trouble-prone, 70-year-old-technology magneto." The auto engines all include modern all-electronic solid-state ignition systems.

My secret? I *like* magnetos! Yes, they're old technology that Glenn Curtis probably laughed at. Yes, electronic ignition allows tailoring the spark to the running condition—retarding spark to help starting, or advancing it for maximum power.

But when you hold a magneto, your hand is cupping your plane's *entire* ignition system. It doesn't depend on a battery. It can't be stopped by a burned-out solenoid or a shorted-out power bus. It's not a plastic box crammed with static-sensitive integrated circuits and connectors with corroding pins. A magneto contains a magnet, a rotor, a coil, a breaker switch (points), and a capacitor, and *that's it* (Fig. 12-17).

Fig. 12-17. The aircraft magneto is old technology, but it is also a compact all-in-one system.

The engineer in me *loves* such simple solutions, and a magneto is as simple as an ignition system can get. Sure, they aren't perfect; that's why the FAA makes us carry two of them, but dual failures are vanishingly rare. If the engine suddenly quits dead, the failure is probably somewhere other than the ignition system.

Of course, if the engine just won't start, there's a good chance the problem is with those dad-blasted magnetos.

Magnetos operate very similarly to a traditional automotive ignition system, with one important difference. A magneto generates all ignition power internally, rather than needing juice from a battery or generator.

Figure 12-18 shows the operation of a basic magneto. The unit bolts onto the accessory case and engages a gear located inside. The internal shaft of the magneto turns three things: magnets, cam, and rotor.

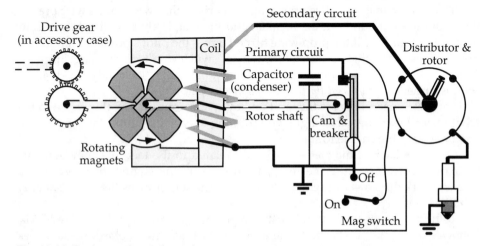

Fig. 12-18. Basic magneto schematic.

The coil is wrapped partially around the rotating magnets. Their rotation induces current flow in the coil, just like any generator. This current flows in the primary circuit, through the closed breaker points, and back to ground. This current through the coil produces a magnetic field. The secondary circuit is affected by the primary's field and builds its own voltage in response; however, its voltage is dependent upon the rate of change of the magnetism of the primary, which doesn't really change very fast.

Rotation of the shaft turns a cam that opens the breaker points and disrupts the flow of current in the primary. Its magnetic field collapses very quickly, and the secondary picks up the change and amplifies it. This high voltage is transmitted to the appropriate plug by the rotor and distributor.

When the breaker points open, the voltage in the primary would normally just arc across the points. That's where the capacitor (or condenser, to a traditionalist) comes in. It acts as a sort of high-voltage temporary battery and momentarily absorbs the voltage. By the time it's fully charged, the points are too wide to arc across. Without the capacitor, spark power is lessened and arcs tend to scorch the breaker points. Without the capacitor, the strength of the spark is severely lessened.

Just like a car, timing is crucial. The magneto produces sparks at intervals; if the cylinder isn't at the appropriate position at the time the plug fires, efficiency is lost. The magneto body is rotated to adjust this timing. You'll note a slotted track for the mag's hold-down bolts. This allows your mechanic to loosen the mag, adjust it to the right timing, and tighten it down in that position.

Timing the magneto is not an owner-maintenance procedure. Bad timing can cause immediate and severe internal damage. Leave it to your A&P.

As a pilot/owner, your main interface with the magneto is turning it on and off. Take a look at the figure again; see how the mag switch is wired. When you flip it to the OFF position, you're actually grounding out the breaker points. The magneto can't fire because (electrically) the points never open.

That's if the mag switch is operating properly. If the switch or its wire breaks, the pilot cannot ground out the breaker, and the mag is always on, which means that a plug could fire if the prop were turned. From the pilot's point of view, this is great because an in-flight failure of the wire or switch won't kill the mag. From the owner's point of view, it stinks because the engine could "kick" anytime that the prop is turned. *Always treat the prop as if a mag is hot.*

For the owner who loses the ignition key at a remote airstrip, things are somewhat rosier. Disconnect the single wire running to each mag from the mag switch and handprop the engine to start.

One drawback of magnetos is directly related to the fact that they generate their own power. Because the mags are turned by the engine, they produce a weaker spark the slower the engine turns. And the starter turns the engine at only 200 RPM or so.

To help, some mags have a booster to generate a hotter spark at slow RPMs. "Shower of Sparks" is the tradename of one such scheme. Retarding the spark (making it later) helps starting at slower speeds. Most airplane engines have at least one mag that retards the spark at low RPMs. You'll hear a "click" or "snap" as the engine is turned over slowly; that's the sound made by the *impulse coupling*.

Like the old-fashioned automotive ignitions that they resemble, magnetos are subject to a number of ills. Breaker points burn, condensers open or short. Coil problems are tricky. Often the coil will work well when cold (or in denser atmosphere), but develop an open circuit when they warm up, or get to altitude. A mag that checks out well during a runup at home might show a miss after a short layover at an intermediate field, and then work perfectly when the mechanic finally arrives to look at it. When you suspect magneto problems, take careful note of the exact conditions: temperature, altitude, time of flight, and the like.

Spark plugs

Proper preparation and cumulative experience changing spark plugs will serve you well in this area of owner maintenance. In addition to the deep socket mentioned in the tool list, you'll need a set of replacement spark plug gaskets. They cost a quarter each, or less if you buy a whole box.

Before removing the plugs, blow the loose dirt and scale away. If you don't have an air compressor, at least puff through a piece of rubber tubing held around the base of the plugs, or use a can of photographer's compressed air. You don't want any loose material to fall into the open hole.

While the general theory of the aircraft ignition system is similar to that of a car, components are a bit different. Everything's a bit more rugged on the airplane. The ignition harness for most airplanes is heavily shielded to prevent radio interference. The harness screws onto the plugs rather than just clipping to them (Fig. 12-19).

To remove a plug, first loosen the harness nut atop the plug. When loose, pull it straight out. The harness terminates in a ceramic cylinder, called the *cigarette*, with a small spring at the end. Look, but don't touch. Finger oils can cause contact problems and misfiring. Examine the spring carefully; if discolored by corrosion, clean it off with sandpaper.

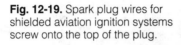

Fig. 12-19. Spark plug wires for shielded aviation ignition systems screw onto the top of the plug.

To take the plug out of the engine, you'll need your deep ⅞" socket. Clip it to your wrench, and slide the socket down the plug until it's well seated.

Plugs are installed to a particular torque. So they'll need a bit of a heave to get them to start turning counterclockwise for removal. Don't just wrap both hands around the end of the wrench and pull, though. The socket is so long that just pulling on the handle can apply some bending load to the plug as well. Instead, support the head of the wrench with one hand while applying torque with the other, as shown in Fig. 12-20.

When the plug loosens, use the ratchet to back it the rest of the way out. If you're going to pull all the plugs at this session, have a muffin pan or egg carton to set them in. Keep each plug separate and label them to indicate which cylinder they came out of, and whether it was the top or bottom plug. If you drop one, throw it away. Unseen damage can rear its head later.

While the layout of the electrodes at the end of each plug is a bit different from a car's plug, the electrodes are the same basic design. The ceramic insulator shouldn't be cracked or have any pieces missing. The electrodes should be intact, not bent or

Wrong

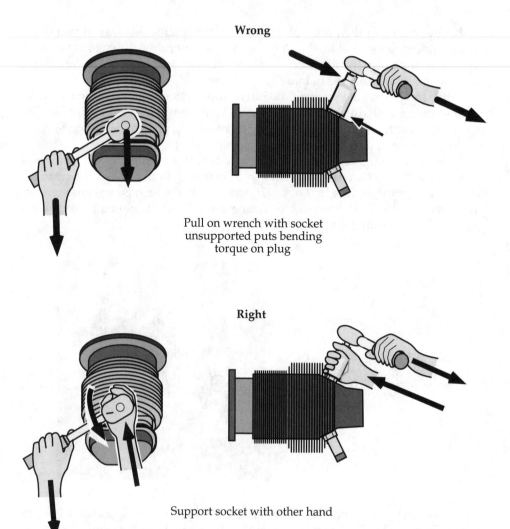

Pull on wrench with socket
unsupported puts bending
torque on plug

Right

Support socket with other hand

Fig. 12-20. Don't damage a spark plug by using only one hand on the wrench. Use the right tools to help you use both hands.

melted. A brownish-gray deposit around the electrodes is the sign of normalcy. The center electrode starts out circular, but goes gradually out of shape with wear.

Black deposits indicate carbon or lead fouling. If the deposits look oily, it could indicate a ring or valve problem in that cylinder. If any plug seems abnormal, it's time to talk to your mechanic. Remember which cylinder(s) the bad ones came from.

One all-too-common problem that you can alleviate is lead fouling. Many engines designed to run on 80 octane fuel have trouble with the lead contained in 100LL. This lead tends to clump in the plug, forming little "nuggets." Carefully scrape them away with a fine tool like a dental pick (a screwdriver is too clumsy) as Fig. 12-21 shows. If you have access to a spark plug cleaner, a little easy blasting should help.

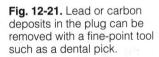

Fig. 12-21. Lead or carbon deposits in the plug can be removed with a fine-point tool such as a dental pick.

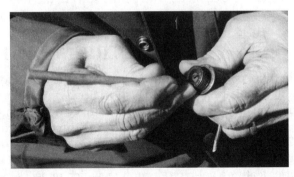

The lead causes more problems than fouled-up spark plugs, of course. Solutions include finding a source of 80 octane aviation fuel, buying an autofuel STC, or using Alcor's TCP in your fuel. TCP doesn't do anything magical to the lead, it just ensures that the lead deposits are softer and helps the engine expel the lead rather than form solid clumps on the inside of the combustion chamber.

Use a soft brush to eliminate any remaining loose particles on the plug prior to setting the gap.

I used to adjust the gap on my '74 Volkswagen by pressing the electrodes down against the fender. Airplanes require a bit more finesse. Pick up an aviation plug gapping tool. Prices vary, but a good-quality gapper can be found for $20 or less. You'll also need a gap feeler gauge, which costs about the same price. Practice a bit on an old plug (or get some "dual instruction" from a mechanic) prior to risking your flight plugs.

Before reinstalling the plugs, you might want to shift them around to equalize wear. Bottom plugs should become top plugs and vice versa. Shift them diagonally between cylinders, too; the top plug on the front leftside cylinder should be moved to the bottom hole in the back rightside cylinder. The engine manual should show the proper rotation.

The reason for shifting from the top to the bottom is pretty obvious. But why switch between cylinders? The magneto is essentially an alternating-current generator. Relative to the aircraft ground, it produces a high-voltage positive pulse, followed by a negative pulse. The way the distributor works, each cylinder gets the same polarity each time. Switching the plugs around changes their firing polarities and evens the wear.

There's one very important last step prior to reinstalling the plugs. They are subjected to vicious extremes of heat and pressure that can tend to lock them in place, which puts you in a world of hurt the next time that you want to inspect the plugs.

To prevent this problem, coat the second and third plug threads with a special lubricant called Anti-Seize. It's available through aviation parts sources. Be careful not to get any on the electrodes, and don't use it on the nut that attaches the harness.

Slide a new gasket onto the plug. Start screwing the plug into the cylinder by hand. Put your socket on a torque wrench and tighten the plug to the torque specified in the service manual, usually about 28–30 ft.-lbs.

Clean the cigarette with a little solvent (acetone or unleaded gas), let dry, and slip it into the plug. Tighten up the nut, and you're done.

ELECTRICAL SYSTEM

For those familiar with car electrical systems, this should be old home week. Those with fumble-fingers should also rejoice because most of the allowed preventative maintenance involves little more than changing a bulb or two.

A simplified diagram of a typical aircraft electrical system is shown in Fig. 12-22. It all starts with the battery. There are several different types. Some batteries are practically indistinguishable from the lead-acid units used in cars. Other batteries feature a gel-type electrolyte, which replaces the liquid acid with a substance that doesn't vent gas when charging. These batteries are completely sealed and require little maintenance.

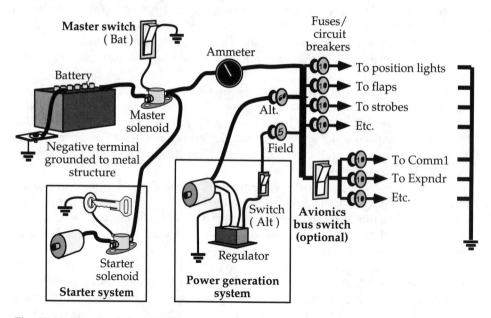

Fig. 12-22. Simplified electrical system schematic.

The negative terminal is attached to the aircraft's metallic structure, *grounded* is the technical term. Most devices on the aircraft can then connect to the negative terminal by simply achieving a good connection to the metal structure.

The positive terminal of the battery goes to one contact of the master solenoid. Turning the master switch on doesn't physically connect the battery to the aircraft power system. That would require a heavy-duty switch in your panel with attendant long runs of heavy battery cable.

Instead, the small master switch in your panel activates the master solenoid, which includes the thick copper plates necessary to work with such current. Other terms for solenoids include "contactor" and "relay." It is the source of the big "Clunk!" you hear when you turn on the master.

The master solenoid brings enormous safety advantages. My dad's car burned up in an electrical fire. If he'd been able to disconnect the battery early, the fire probably wouldn't have happened.

That's what the master solenoid is all about. Turning off the master switch disconnects the battery from most of the airplane. Exceptions include the clock, which is protected by its own fuse.

Two heavy cables emerge from the outlet side of the master solenoid. One goes directly to another solenoid that is for the starter. On most airplanes, the key activates the starter solenoid, which then applies power to the starter motor itself.

The other cable goes to the main electrical power bus and to the aircraft's generator or alternator system. The power for individual electrically driven components is run through a fuse or circuit breaker for protection. On some airplanes, a separate switch is provided for an avionics power bus. This protects sensitive electronics from the power variations produced during starting or stopping the engine.

To provide power while the engine is running, the aircraft includes either a generator or an alternator. The difference isn't really important, but most modern planes have an alternator because they are better at producing power at low RPMs, don't have heavy current brushes and commutators to produce radio noise, and are much lighter.

Electrical woes

Aircraft electrical systems seem to be involved in more than their share of problems. Let's take a look at some of the causes and what you can do to mitigate them.

Let's start with the battery. As mentioned earlier, the negative terminal is attached to the aircraft structure. This avoids having to run *two* wires to all the electrical devices. Each device becomes grounded upon solid connection to the metal structure, perhaps as a result of installation, perhaps with a ground wire.

While simple and lightweight, grounding in this fashion can lead to difficulties. Engines are usually mounted on insulating rubber shock mounts; hence, you'll see a *ground strap* running to the engine block to ensure its grounding. Other areas might feature such straps as well; the bare braided metal that they are made of is easily recognizable.

Loose ground straps can cause irritating electrical problems. Components won't work, they will flicker, and they will develop strange anomalies.

If weird electrical problems arise, start checking the ground straps. Check that they're solidly bolted in place and that the braid is solidly connected to the terminal. You might even unbolt them and clean the contact surfaces with a file or emery cloth. This isn't included on the list of preventative maintenance, but as long as it's a simple connection, the FAA shouldn't mind. If you are concerned, check with an A&P.

The master and starter solenoids are also sources for recurring problems. When you turn the switch on, you're applying power to an electromagnet. As shown in Fig. 12-23, this pulls in a soft iron core that is attached to a copper bus bar, which connects the input and output of the solenoid. A break in the rather fine wire of the electromagnet's coil will keep the solenoid from activating. And while the solenoids are usually sealed, corrosion still seems to appear. It can prevent the contacts from making good contact, or can jam the core so that it can't move.

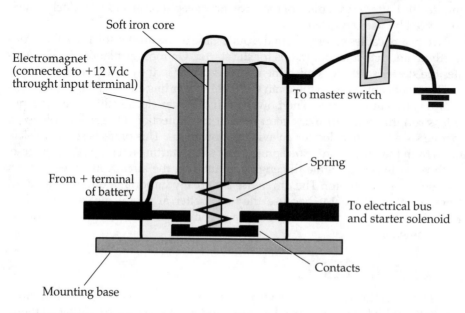

Fig. 12-23. Workings of the aircraft solenoid.

Unfortunately, solenoids can also fail in the *inverse* manner; they can activate and refuse to turn off. If this happens on the starter solenoid, you can be in deep trouble and not even know it. The engine's running, but the solenoid keeps the starter gear engaged and tries to keep turning the starter. If you don't notice, it can quickly chew up the starter bendix and starter ring gear. On some engines, shavings from the bendix and ring gear will fall into the oil flow, which then spreads this abrasive poison through the engine—deep trouble, like I said.

You might be able to hear it as a sort of rattle from up front. Depending upon the electrical system, the ammeter could provide another clue. Look back at Fig. 12-22. The ammeter isn't positioned to tell the pilot what the starter's current draw is. But if the starter keeps turning, the generation system will try to supply power for it. This will show up on the ammeter as a high rate of *charge*. The alternator thinks that it's charging the battery, when in reality it's powering the starter.

Indications might be different on your airplane's engine, though. Talk to your mechanic. The important thing is to shut down the engine and switch off the master if you suspect the starter is still powered.

Problems with the master solenoid are minor in comparison, but still irritating. If the master solenoid stays on, it'll drain the battery. If the battery is fully charged but the solenoid won't activate, you still don't have power.

There is even the potential for combining these problems. Corrosion can keep a solenoid from working; it can also provide a high-resistance path for slow current flow, even when the master switch is off. If you fly the plane every day, you don't notice anything wrong. But if you only get out to the airport every two weeks or so, the battery has enough time to discharge through the high-resistance path. The battery is dead.

Often, your immediate reaction is to replace the battery for about $130, but the problem is *still* there. I know. I went this route. Before making the same mistake I did, check out the solenoid. If you've got an multimeter, disconnect the battery and measure the resistance across the solenoid terminals. It should be infinite.

Without a multimeter, charge up the battery and disconnect it from the aircraft before you leave. If the battery still has enough power the next time you come out to fly, the problem probably lies in your master solenoid. If the battery is flat, it is time to install a replacement.

Battery maintenance

Maintaining your aircraft battery is fairly simple. After removing the battery-case cover (if any), you'll be facing the top of the battery (Fig. 12-24). The aircraft cables bolt to the battery rather than clamp around terminals.

If you've got a conventional lead-acid battery, unscrew the caps and check the electrolyte level. The liquid should be at the base of the rings. If the level is low, add distilled water to bring it up to the mark.

Replace the caps. Take a wad of paper towels, sprinkle baking soda on them, and wipe the liquid away from the top of the battery. Yeah, it's probably just spilled water. Use the baking soda anyway to neutralize any acid that might be present. Make sure the battery caps are already on. (Always wear old clothes that can be thrown away if damaged by battery acid.)

Corrosion can work its ugly way between the terminals and the power cables of any type of battery. You should disconnect the cables on occasion and service the terminals.

Take a look at the terminals. The positive should be marked with a plus (+) and the negative one with a minus (–). Disconnect the *negative* terminal first. Remember, it's grounded to the aircraft; every piece of metal on the plane is electrically the equivalent of the minus side of the battery. If you tried to disconnect the positive terminal first and your wrench slipped and touched metal, you'd throw one heck of a spark: not good for the battery, not good for the wrench, not good for the metal the wrench is touching, and quite possibly not good for *you*.

Fig. 12-24. Most aircraft batteries are contained in boxes under the cowling. Note the lift strap for removal.

When the negative terminal is undone, position the cable aside so that it can't accidentally remake contact with its terminal on the battery. Then disconnect the positive cable.

At this point, you might as well pull the battery out of the plane. There might be some holddowns to unbolt. When free, the battery should lift right out. Clean the battery terminals and cable ends with a stiff metal brush that is commonly found at auto-parts stores. If you prefer, look around for a special terminal and cable cleaning tool.

Take a look inside the battery box. Is it a little wet in there? Dump some baking soda inside and flush it out with fresh water. Wipe it down with soda-sprinkled paper towels.

Lift the battery back into position. Wipe Vaseline liberally over the terminals and the aircraft's cable lugs. It doesn't interfere with power transfer, but it does slow corrosion. Connect the *positive* terminal first and verify the polarity match and tightness. Connect the *negative* terminal second and verify polarity and tightness. Close up the battery box, and you're done.

Bulb replacement

According to FAR 43, the owner is allowed to replace ". . . bulbs, reflectors, and lenses of position and landing lights." Easy enough: Usually it's just a few minutes work to remove the lens or cover with a screwdriver (Fig. 12-25), swap the old bulb for a new one, and replace the cover.

Watch out for a couple of "gotchas." First, before putting the new bulb in, take a look at the socket. Sometimes a socket will get wet and a bit rusty; if so, dry out the socket and clean up the contacts. Shoot in some WD-40 prior to inserting the new bulb for some protection.

Second, landing lights and some position lights are covered by Plexiglas lenses. Plexiglas stretches and shrinks with temperature changes; well, so does everything, but plexi does it a lot more than aluminum. Because the lenses bolt into aluminum housings, the screw holes in the Plexiglas are a bit larger than they have to be, which permits movement with the temperature changes.

If you tighten the attachment screws too much, it won't let the Plexiglas move. A few weeks or months later, you'll find cracks running between the mounting holes, so don't tighten them all the way. The screws generally go into self-locking anchor nuts; therefore, they aren't dependent upon mounting torque to hold them in place. Leave the Plexiglas a tad loose.

Finally, reports indicate that the orientation of landing light bulb filaments can have an effect on the bulb's longevity. When possible, orient the filament vertically instead of horizontally. If bulbs still tend to burn out quickly, experiment with various orientations.

FOR FURTHER INFORMATION

There are several sources of detailed information on aircraft systems and maintenance. TAB's *Aircraft Systems: Understanding Your Aircraft*, by David A. Lombardo, is an excellent reference and belongs on every aircraft-owner's bookshelf.

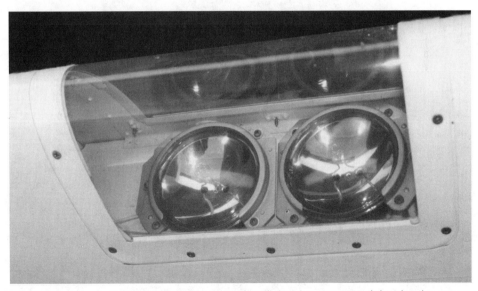

Fig. 12-25. It's easy to replace landing and position lights; however, overtightening the screws can cause a light's Plexiglas cover to crack.

Belvoir Publications offers two magazines of interest to the aircraft owner. The first, *The Aviation Consumer,* has been mentioned before. It contains a lot of mechanical information of use to any pilot.

The other magazine is aimed at professional mechanics and owners who are seriously interested in maintaining their planes. *Light Plane Maintenance* provides in-depth articles on aircraft operations and maintenance issues. LPM takes an aggressive stance and is entirely supported by subscribers (i.e., no advertisements).

For subscription information about either magazine, contact:

Belvoir Publications, Inc.
P.O. Box 420234
Palm Coast, FL 32142
800-829-9085

Information published in LPM has been compiled in Belvoir Publication's *Light Plane Maintenance: Aircraft Engine Operating Guide,* by Kas Thomas. It is aimed at pilots of all skill and experience levels. Articles address problems and procedures applicable to all piston-aircraft engines.

Kas Thomas also publishes a bimonthly engine-technology journal for pilots. For more information about *TBO Advisor,* write the journal at P.O. Box 625, Old Greenwich, CT 06870.

A highly recommended reference for those serious about engine operation is the *Sky Ranch Engineering Manual,* by John Schwaner. It provides excellent explanations of how failures occur, what the operator can do to minimize them, and what is involved with the repairs.

13

Problems

IT WOULD BE NICE IF your airplane ownership went smoothly, with nothing to mar the joy of flight. That's the way it goes for most aircraft owners, but there's always that niggling little worry. What if it breaks down at some little field somewhere? How do I get the plane back if I crash-land somewhere (Fig. 13-1)? This chapter discusses some of these potential problems. Some situations are minor; some aren't.

Fig. 13-1. How are you going to get your plane back to the airport if you land off-field? This transportation method is limited to short distances only!

DEAD BATTERIES

Considering all the other calamities that might befall you, finding a dead battery in your plane is pretty tame. Chapter 12 reviewed the electrical system and how a faulty master solenoid can drain the battery. Of course, sometimes the battery is dead for a less-exotic reason: You left the master switch on.

Finding the problem

If the master wasn't left on, two possible sources of the problem are the battery or the master solenoid. A minute or two of troubleshooting is called for.

Listen when you turn on the master switch. Do you hear the familiar "clunk" of the solenoid activating? If so, there's a good chance the problem lies in the contacts of the solenoid rather than in the battery. To make sure, turn on the cabin light; if you hear the solenoid activate, but there's not even a dim glow from the bulb, the solenoid is probably bad.

Find a mechanic to install a new solenoid. Yes, the engine *can* be started by handpropping, but alternators require an external power source to "excite" their field windings at startup. If the battery isn't there, the alternator won't work and could be damaged if the engine is started. Generators also require an external source when first started, though they get by on less.

All right, but what if you *don't* hear the solenoid activate when the master switch is turned on? Is the battery discharged or is the solenoid bad?

The best solution is to open the battery case and measure the battery voltage. Radio Shack sells several kinds of *multimeters* that measure voltage, current, and resistance. Resist buying the larger, more expensive ones, unless you can use them in other electrical work. For troubleshooting aircraft batteries, the cheap models are adequate.

Set your meter to the lowest range that will measure your battery's nominal voltage (i.e., the "0–15 volt" or "0–20 volt" range for a 12-volt battery). Touch the red multimeter probe to the battery's positive (+) terminal, and the black probe to the negative (–) terminal. If the battery is fully charged, it should read at least its rated voltage; any less is cause for suspicion.

If it's *apparently* OK, leave the red probe on the positive terminal and touch the black probe to a bit of exposed metal structure. If it doesn't read the same voltage, the ground cable from the battery to the aircraft structure has a poor connection.

If the ground is *apparently* OK, move the red probe to the input part of the solenoid (you might have to peel back a rubber boot), keeping the black probe touching the metal. If you still get voltage, the problem lies in the solenoid. Or just perhaps in the master switch or its wire to the solenoid. Turn on the master switch, hold the red probe to the positive terminal of the battery, and touch the black probe to the small connector on the solenoid. If you read battery voltage, the switch and wire are good and the solenoid needs replacement.

If the problem is the battery?

Charging a dead battery

The best solution to dead-battery woes is to remove the battery and charge it. Disconnect the battery as described in chapter 12. Remove it from the airplane and bring it to the charger. Don't place the battery directly on a metal or concrete surface that will act as a heat sink.

Remove the battery caps. If it's a lead-acid battery, check the fluid level and add distilled water if necessary. Then connect it to the charger: red clamp to the positive (+) terminal, black to the negative (–). Turn on the charger.

The meter on the charger should show a fairly high rate of charge. Go away for a few hours. When the charge rate has dropped to a moderately low value, disconnect the battery, replace the caps, and reinstall it in the aircraft as described in chapter 12.

Jump-starting

If the battery isn't completely dead, but just can't seem to turn the prop over, there are a couple of other options.

As long as the battery still has some charge, handpropping can get you on your way. Remember, it needs that residual charge to excite the alternator or generator. If the battery is completely flat, handpropping won't do it for you.

Plus, if your plane mounts a Lycoming and you've already tried to start it, hand-propping will be a bear. The starter bendix stays engaged. When you prop the engine, you'll also be turning the starter through the ring gear and bendix. Maybe it doesn't add that much more friction, but the grinding sounds horrible.

Whether the engine is a Lycoming or Continental, *don't even think of trying to handprop it unless you've had training*. Instead, jump-start the airplane just like you would a car. In fact, if your plane has a 12-volt electrical system, you can jump it *from* your car. Some planes make this easy with external power receptacles. Otherwise, you'll have to hook up a set of conventional jumper cables.

Either way, start by removing your plane's battery caps. If you're working alone, make sure the plane is tied down, the chocks are in place, and the parking brake is set. If someone is helping you, make every effort to ensure the helper's safety.

If the plane has a power receptacle, you'll need a special set of jumper cables. These are pretty expensive, almost $100. All you have to do is hook up the cables to your car's battery and plug in the connector.

Without a receptacle, things get a bit hairier. After the engine starts, you'd like to just disconnect the cables and taxi away. Some planes have their battery in the baggage compartment. That's not so bad, though it can be awkward to get to.

On most, it's forward of the firewall. This gets a little sporty. When the engine starts, you'll be working pretty close to the rotating prop to get things buttoned up again. Hopefully you'll be able to get at the battery though some small access door. If you have to remove the top cowl to get to the battery, count on running the engine for a while to charge up the battery, then shutting down to reinstall the cowl.

When you have determined a safe procedure to follow after the airplane engine starts, you may proceed with the jump-start. Connect the positive cable (red) to your car's battery, then to the positive terminal on your aircraft. Connect the negative terminal (black) to the aircraft's battery, then to a grounded point under your car's hood. Run your car at a higher RPM for a while to put a little power into your aircraft's battery, then go ahead and start the airplane.

Jump-starting isn't really the best thing for your airplane. The alternator will stay healthy longer if it doesn't have to provide the massive charge that a flat battery demands. Jump-start as little as possible.

STORAGE

There comes a time for some owners when the airplane can't be flown for a while. Maybe you live in an area where bad weather shuts down flying for months on end. Perhaps you've developed a temporary medical condition which will prevent you from flying for a bit. Or maybe the money just isn't there.

Inactivity is hard on airplanes; tires stiffen and engines rust internally. You do an airplane no favors by letting it "rest." It is going to deteriorate on you anyway. Fly it, if possible. Otherwise, there are a few precautions you can take, depending upon how long the plane's going to be laid up.

Short-term inactivity

If you won't be flying for just a month or so, there are a few things to do. First, fill the fuel tanks. Rubber fuel tank bladders dislike exposure to air. Remove the battery and take it home. Put it on a trickle charger one night a week. Put some desiccant in the cabin to reduce dampness. Hardware stores carry products to reduce humidity in storage areas. One brand is Dry-Z-Air.

Cover all ports and inlets to keep the bugs, birds, and rodents out. If mice flourish in your area, consider putting metal shields around the tires to keep them from climbing the landing gear into the wings or cabin. Set some traps inside and outside.

Spray exposed hinges and linkages with a heavy corrosion-preventative oil. Your mechanic can probably recommend a particular brand, I use one called LPS 3. Spray brake disks with light oil; remember to wipe it off before the next flight. Wipe preservative on all exposed rubber parts.

Every week or two, take your battery out to the airport and hook it up. Turn on the strobe and radios. Tune the radio to all its frequencies and turn the volume back and forth to wipe off any corrosion on the contacts and wipers. Warm the radios for at least 15 minutes.

Roll the plane forward and backward a few times to exercise the wheel bearings. Don't put it back in exactly the same position; set it slightly differently so another part of the tire is touching the ground. If all else fails, bring your jack, lift each tire a bit, and spin them for a minute or so.

Turn on the fuel valve and exercise it to all positions. If you've got a boost pump, switch it on for a second. Rubber parts in the carburetor can tend to dry out, too.

Work the throttle, mixture, and carb heat controls back and forth. Lower and raise the flaps. Move the control wheel or stick gently to its limits a few times. Check the preservative oil you've sprayed on the hinges and renew if necessary.

Making sure the magneto switch is off, the chocks are set, and the tiedowns are tight, step to the front of the airplane and turn the engine over by hand. Do at least 10 blades (five complete turns of the engine).

Do *not* start the engine and idle it. Combustion byproducts include water vapor, which reacts with the other elements to form acids and other nasty stuff. Collection of these harmful materials is halted by getting the engine warm enough to drive the moisture out. The engine won't get warm enough during a ground runup.

Turning it over by hand wipes any condensed moisture off the cylinder walls, slowing the advance of corrosion. Do this as often as possible; engine manufacturers generally recommend turning an inactive engine by hand at least once a week.

Your plane should be able to weather a month or two using these procedures—perhaps more than two months during cold weather. It's not *good* for the engine, but no serious harm will likely occur. If you can still fly it occasionally, the storage period can stretch out.

To be sure, get advice from your mechanic.

Long-duration storage

You say that you are not going to be able to fly your plane for six months or a year? Sell it. Long-duration storage is hard on your airplane. It might never make it back to the level of reliability it had. Corrosion takes its toll, both on the airframe and inside the engine. Avionics connectors develop internal oxidation and strange problems result. Grease runs out of wheel bearings, control pulleys, and propeller hubs. You're better off in the long run to cut your losses and sell. Buy another plane when you're able to fly again.

Oh, all right. Selling is not always an option. Sometimes the plane fits your needs perfectly; you're loath to get rid of it. Maybe it's a rare model. Admit it, sometimes we get emotionally attached to our metal birds. Get with your mechanic and plot out the storage plan. He or she will probably recommend the following course:

- Draining all the engine oil and replacing it with preservative oil.
- Spraying the inside of the cylinders with hot preservative oil.
- Replacing the spark plugs with special plugs containing desiccant (Fig. 13-2).
- Blocking the exhaust pipes with desiccant plugs.
- Spraying the inside of the airframe with a corrosion fighter such as ACF-50.
- Putting the airplane up on blocks to preserve the tires.
- Taping/sealing all gaps.

Storage location is important, too. In many parts of the country, outside long-term storage is not practicable. You'll need a roof over its figurative head.

If you've got a good-size garage, you could take the plane apart and store it there. But if you ever decide to sell it, you'll take a real loss. Flyable airplanes are worth much more than disassembled "projects."

Fig. 13-2. Clear-plastic desiccant plugs contain crystals that absorb moisture from within the engine. The crystals change color to indicate when they must be replaced.

Taking it out of storage is a major undertaking as well. In all probability, it'll be due for an annual, anyway. A lot of bad things can happen if a plane sits inactive for too long. Brace yourself for a steep bill.

PROBLEMS AWAY FROM HOME

Few things are as stressful as having your plane break down on a trip (Fig. 13-3).

Fig. 13-3. Stan Brown contemplates a solenoid failure on *Ramp Rooster* (see chapter 4).

Tools

Consider stowing a few tools in the baggage compartment, nothing heavy or large, just some stuff that might just get you going if something minor happens. This assumes that you are accustomed to maintaining your own plane; don't try to learn on a windswept ramp 200 miles from home.

You can usually get by with a small tool box, or perhaps a cloth bag. Put a ⅜"-drive socket set in it (hardware stores sell inexpensive socket sets that are suitable for occasional work), some screwdrivers (with a variety of tips), a set of common wrenches (⅜", ½", ⁹⁄₁₆", ⅝", and ¾"), a couple of types of pliers, some safety wire, wire twisters, and the like. Add a sprinkling of extra aviation-quality nuts, washers, and cotter pins.

Keep your spark-plug socket in the box, plus some extra copper washers. Save a couple of old plugs from the first time you install new ones. Wrap them carefully, stuff them inside a small cardboard box, and tape the box shut.

If you ever replace an inner tube, compress the old one and store it in the far corner of the baggage compartment, *especially if your plane uses an unusual tube*. I got a flat tire at a remote grass strip once (Fig. 13-4). I took the wheel off (using willing bystanders as a jack), and I had all the tools to separate the wheel. Unfortunately no one on the field had the right-size tube.

Keep the number of onboard tools within reason, of course. Most of the time, fixing it yourself is just wishful thinking.

Finding help

If the problem is beyond your skills and knowledge, an A&P will be necessary. At most airports, a mechanic should be easy to find during working hours on a

Fig. 13-4. The Fly Baby's left tire went flat at this remote strip during a picnic. There were plenty of people to lift the airplane and tools to remove the wheel. Unfortunately there was no replacement tube available. The plane was secured, and the author caught a ride home to buy a new tube.

weekday. He or she can usually slip a transient aircraft into the schedule to at least diagnose a problem.

On weekends, most FBOs and flight schools have an A&P on-call. Unfortunately, the on-call rate is higher. You might end up shelling out $100 an hour. Freelance A&Ps might be a better pick. Wander around the airport, and talk to some owners. Explain your problem and ask for an A&P recommendation. Freelancers often work on weekends.

Ferry permits

Sometimes you'll have a problem that doesn't make an airplane physically unflyable, yet is not legally airworthy. The gear-retraction system develops a problem, for instance, or you get weathered-in at a remote airstrip and the annual expires. Because the airplane does not then correspond to the requirements of its type certificate, its airworthiness certificate is invalid until the fault is corrected.

The FAA can authorize flight under these conditions using a "Special Flight Permit," commonly known as a *ferry permit*, which is covered under FAR 21.197:

> (A) A special flight permit may be issued for an aircraft that may not currently meet applicable airworthiness requirements but is capable of safe flight, for the following purposes:
> (1) Flying the aircraft to a base where repairs, alterations, or maintenance can be performed, or to a point of storage.

A special flight permit actually is a temporary replacement for your aircraft's airworthiness certificate. To obtain the permit, you will use FAA Form 8130-6, which is the application for an airworthiness certificate.

Apply via an FAA flight standards district office either in person or by phone. Have all your paperwork ready and be prepared to explain the situation to the inspector. If he or she approves your application, you'll receive a temporary airworthiness certificate. If you've applied by phone, the FAA will fax it to you. *This form must be kept in the aircraft*.

Prior to the flight, an A&P must examine the aircraft and certify that the plane is safe for the proposed flight. The FAA might require certain additional steps be taken. For instance, perhaps a retractable aircraft has a landing-gear problem. The FAA will authorize you to fly the aircraft with the gear locked down. The agency will also require that the pilot be *unable to retract the gear*. The A&P will have to disconnect the switch or take some other means to prevent retraction.

Be aware of several points related to a special flight permit:

- The special flight permit exists to allow flight to a location where maintenance can be performed, not to let you fly to a place where you can get it done cheaper. If there are extenuating circumstances, such as wishing to fly the plane to a specialist for your aircraft model, bring them to the attention of the inspector.
- The permit only authorizes solo flight. If you're on a family jaunt, all passengers will have to get home another way.

- The permit authorizes *one flight*. When you reach the destination, the permit expires.
- Check your insurance because it might not cover your aircraft when operated under a special flight permit.

OVERHAULS

"The airplane needs an overhaul."

Ugh. That's probably the worst thing you can say to an airplane owner. I can't give you many words of comfort here. There's no question that it's going to cost big bucks. It's not that you don't have any options, it's just that you don't have any low-cost options.

When do you overhaul the engine? Recall from previous discussions that the engine's TBO has no legal significance in noncommercial operation. As a private operator, you can legally fly it until it comes apart; however, it's best to overhaul before that point.

A number of "triggers" commonly initiate the overhaul process. The first, of course, is catastrophic failure. A connecting rod jutting from the crankcase is Mr. Lycoming's way of saying, "Get out the checkbook." It doesn't have to be quite that spectacular. The sudden appearance of large pieces of metal in the oil is another sign.

Sudden catastrophic failures without warning are rare. In most cases, warning signs just aren't picked up. Detecting them might prevent a forced landing, but won't necessarily delay the inevitable overhaul.

Another initiator is a prop strike. Because the engine manufacturer probably calls for a complete teardown and inspection anyway, you might as well spring for an overhaul.

Probably the most common trigger for an overhaul is lack of power. Over years of operation, springs get weaker, bearings wear down, and tight fits become sloppy. The engine just gets tired.

Cylinder compression is one of the primary indicators of engine condition. While conducting an annual, a mechanic uses compressed air and pressure gauges to measure how "leaky" each cylinder is. As Fig. 13-5 shows, air regulated at 80 PSI is applied to a calibrated ¼"-long 0.040"-diameter orifice, with the outlet connected to the cylinder via a spark plug hole. The air pressure is measured on each side of the orifice. The cylinder's piston is at top-dead-center (TDC) of the compression stroke—all the way to the top with intake and exhaust valves closed.

If there's absolutely no leakage in the cylinder, the pressure gauge on the cylinder side of the orifice indicates 80 PSI. But if the piston rings don't quite seal right, or either valve isn't seated solidly, some air will leak out. The *compression* of that cylinder is the ratio of the input pressure vs. the pressure inside the cylinder; 80/65 means that 80 PSI was applied to the upstream side of the orifice, but enough air leaked through from the cylinder to reduce the pressure by 15 PSI.

Where that 15 PSI went is informative. Your A&P will use his or her ears. If the carburetor's hissing, the problem is probably with the intake valve. At the exhaust

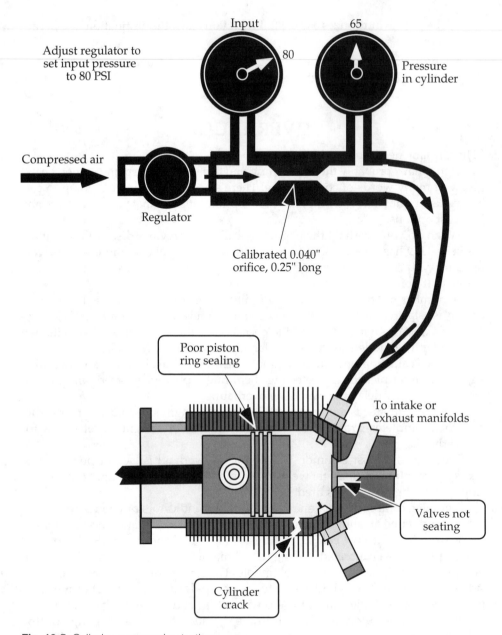

Fig. 13-5. Cylinder compression testing.

pipe? Exhaust valve, of course. If the hissing comes from the crankcase breather, the rings are most likely the culprit.

A cylinder with 80/65 compression is on the low side, but acceptable. It could fly for years at this level. Compression will vary up and down, depending upon a number of factors. Normally you'll see a gradual trend downward.

Compression lower than 80/60 usually grounds the engine. If one cylinder is low, but the others are acceptable, you might be able to get by with just repairing the offending cylinder. Sometimes the only problem is a little carbon around the valves.

If compression is low in every cylinder, and if the engine is still far from its nominal TBO time, a *top overhaul* (Fig. 13-6) is a cheaper alternative. In a top overhaul, the cylinders, valves, and pistons receive the full treatment, but the crankcase is left undisturbed. The engine doesn't even have to be removed from the aircraft. If the engine has a lot of time on it, you might as well bite the bullet and get a full overhaul.

Fig. 13-6. Top overhaul in progress. The engine can remain in the airplane while the cylinder heads, cylinders, and pistons are removed for replacement and/or refurbishment.

Finally, be aware that compression checks aren't perfect. A little crud in that 0.040" orifice or an improperly regulator will make the cylinders read artificially low. If the cylinder isn't right at TDC, one or the other valve might be slightly open. And each piston ring includes a little gap; these gaps rotate slowly around the piston as the engine runs. These gaps can sometimes line up and cause a temporary compression loss. Don't get railroaded into an early overhaul! Take the plane to another mechanic for a second opinion.

Who's going to do it?

Probably the biggest factor in how much an overhaul costs is in who performs it. Let's look at some options, from the least expensive to the most expensive.

Doing it yourself
Like any other aircraft maintenance item, you can overhaul your own engine as long as a licensed A&P supervises and signs off your work.

If you've got an extensive mechanical background and an A&P willing to take the risk on you, this might be the option for you. Smaller aircraft engines aren't that complex; a Continental A-65 and a VW engine offer about the same minimal complexity. If you can hot-rod a Chevy V-8, a Lycoming O-235 should hold no surprises. You'll be left with a healthier bank account plus a full understanding of the engine's condition.

Unfortunately it is not the solution for most of us. Too much can go wrong. Unless you have the right background, it would be best to leave the engine to a pro.

Freelance mechanic

The next step up is to find an A&P willing to overhaul the engine for you. This is probably the lowest-cost realistic solution to your problem.

Unfortunately it is also filled with risk. How good is the mechanic? Will he or she cut corners to maximize profit? After all, your only way to judge is by how good the engine *looks* afterward. Paint is a lot cheaper than gaskets, seals, and pistons.

My friend Kirk hired a freelancer to overhaul the engine for his homebuilt. Kirk accidentally damaged the engine during reinstallation and took it to another shop for repair. The new shop refused to sign off the engine. Many parts were beyond tolerance; the interior of the engine was filthy. Kirk paid $3,000 more to get the engine into legal condition.

Still, a freelancer is the cheapest route. If you think that you have found the right mechanic, ask for references. Talk to those people who are flying engines that the A&P overhauled. Find out how the engines are holding up.

FBO overhaul

FBO overhauls will cost more because they've got more overhead. Yet the quality can vary, just like a freelancer. Nominally, at least, their shop might be better equipped than someone working in their garage. You might prefer to deal with an established business rather than a private individual. Realize that FBO overhauls can also suffer from a desire to maximize the bottom line.

Repair-station overhauls

Some companies have specialized in engine overhauls. These outfits stake their reputation on their work; hence, you'll see a uniformly high level of quality.

You'll usually see them listed as "Certified Repair Stations." These are governed by FAR 145, which places stringent requirements on facilities, equipment, personnel, and recordkeeping. Companies can gain certification in a number of fields, such as engines, instruments, propellers, or airframes.

The FAA keeps a tight rein on repair stations. They are certified to perform work on rather limited aspects of aviation: "Reciprocating engines of 400 HP or less" or "Instruments/mechanical." Stations can gain ratings in more than one category.

You can expect the highest-quality work from a certified repair station, as well as the highest prices. Yet it can pay off when the time comes to sell the airplane. Listing the engine as "400 SMOH by Victor" has a certain attraction.

One strange aspect of having an overhaul done by a certified repair station: Your engine probably won't be overhauled by an A&P! The airframe and powerplant rating requires expertise in a wide range of skills, from engines to fabric to riveting. An employee of a repair station can gain a *repairman certificate* (or certificates) that allow him or her to perform specified functions while under the station's employ. And because they can specialize in one area rather than many, they can become very good at it.

Factory overhauls

The engine manufacturers have gotten into the act by offering *service-limit* and *new-limit* overhauls, depending upon the engine type. (Service limits are explained in chapter 6.)

A new-limit overhaul by the original builder of the engine is called a *remanufacture*. That again has a certain cachet; a certified repair station can perform the same overhaul, but is not legally authorized to call it a "remanufacture."

Factory overhauls feature their own little bizarre twist: By FAR 43.3, the factory overhaul doesn't have to be performed by a person with an A&P or repairman certificate, and the factory isn't required to be a certified repair station.

Other aspects

Many engine accessories must also be overhauled at particular intervals. Depending upon each accessory's status, you may choose to have them overhauled at the same time as the engine. If you can afford it, go ahead because it will probably be a lot less trouble in the long run.

Many overhaulers don't actually overhaul the engine that came out of your airplane. They give you an overhauled engine in exchange for money and your old engine, called a *core*. If your old engine has significant flaws (a cracked case or a bent and broken crankshaft), there will be additional charges.

Also, if you buy an overhauled engine from a nonlocal company, you'll pay more than just the overhaul cost. A mechanic will have to remove your old engine and crate it. Shipping charges will run several hundred dollars, if not more. When the new engine arrives, the mechanic has to install it and check it out (Fig. 13-7).

This process might take a while. You'll probably want to prepare the airframe for long-term storage without the heavy engine. If the plane is equipped with tricycle gear, taking the engine out will shift the CG back so far that the plane will want to sit on its tail. You'll either need to put something underneath the back end or, like the owner in Fig. 13-8, put some heavy ballast in the engine compartment.

Because the major overhaulers swap a rebuilt engine for your core, your plane might not have to stagnate. Some companies offer complete engine replacement services. If the engine is still safely flyable, you can fly into the site, hand over the plane, and pick it up a day or so later with the brand-new engine already installed.

Alternatives

Rather than facing an overhaul on your engine, there are a couple of other routes you could take. What about a used engine? They turn up on the markets occasionally. Hurricane Andrew destroyed a lot of airframes; yet the engines were

Fig. 13-7. Installing the engine on an Aeronca Champ.

Fig. 13-8. This block of concrete holds the plane in normal ground attitude until the overhauled engine is installed.

generally intact. The insurance companies (or owners of airplanes without hull coverage) put the engines up for sale.

There are several drawbacks to buying a used engine. First, the engine might have been inactive for quite a while. You'll end up paying for your A&P to verify the condition of the engine.

Second, it might be difficult to find the exact model of engine you need on the used market. If your aircraft uses a Lycoming O-320H2AD, then you can't install just any O-320. It must be either an O-320H2AD, or you must go through the STC process to gain approval.

Finally, the competition's stiff. The O-320s used in Cessna 172s, for instance, are rabidly sought by pilots building RV-4 and RV-6 homebuilts.

If you're less than enamored by the performance of your aircraft, overhaul time is the perfect opportunity to solve your problem. If you're going through all the work to remove, overhaul, and replace an engine, why not upgrade to a bigger one? A number of commercial STCs are available to allow installation of larger engines on the common aircraft of the general aviation fleet.

INCIDENTS AND ACCIDENTS

"Stuff happens." What do you do if it happens to you? No one likes the thought of an accident or forced landing (Fig. 13-9). Let's take a few moments to consider some of the factors.

Fig. 13-9. This Cessna 175 set down in a farmer's field when a clogged fuel line starved the engine. Note the damage to the spinner, nosewheel pant, and strut caused when the plane went through a fence.

Primary consideration

The primary thing you should be worried about is medical attention for any injured parties. All else pales in comparison.

Reporting requirements

Aircraft accident investigation is handled by the National Transportation Safety Board (NTSB). The FAA will get involved, but the NTSB is in charge. Accident reporting requirements are contained in the NTSB's Part 830.

If the aircraft receives substantial damage or any person suffers serious injury or dies, the event must be reported to the NTSB "by the most expeditious means possible."

As always, definitions are important. "Serious injury" is defined as broken bones other than fingers, toes, and noses (ouch!), severe bleeding, nerve, muscle, or tendon damage, internal organ damage, second- or third-degree burns, or any injury that requires hospitalization for more than 48 hours.

The NTSB defines "substantial damage" to include damage or failure that affects the structural strength, performance, or flight characteristics of the aircraft, and would require major repair or replacement.

What is more interesting is what the NTSB does *not* consider substantial damage: engine failures, bent fairings or cowlings, dented skin, small punctures in skin or fabric, ground damage to a prop or rotor blades, and damage to landing gear, wheels, tires, flaps, engine accessories, brakes, or wingtips. These events qualify as *incidents*. NTSB's incident definition: "An occurrence other than an accident associated with the operation of an aircraft, which affects or could affect the safety of operations."

Why is this important? Simple. Aircraft accidents are reportable. Incidents, with certain exceptions, are not reportable. If you ding a wingtip on a post while taxiing and fracture the fiberglass, you don't have to tell the NTSB or the FAA. If your engine quits and you safely land in a field, the NTSB doesn't want to know about it.

The exceptions are pretty limited. You must inform the NTSB of any incident involving a flight control system malfunction or failure, any in-flight fire, and any midair collision (consider yourself very fortunate if you are able to file a report regarding a midair). Additionally, any incident where the loss of property (other than the aircraft) exceeds $325,000, and if anyone suffers serious injury or death.

Of course you'll have to report any accidents and probably most incidents to your insurance company.

Securing the aircraft

If you've had an accident, the NTSB requires that you preserve the wreckage. You can't start hauling pieces away, and the general public should be prevented from doing the same. Report the accident to your insurance company as soon as possible; it should handle the security arrangements. If you've set down in a field somewhere, use your tiedown kit to secure the plane until it can be recovered.

Recovery

Whether you've had an accident or just force-landed off-field, the plane needs to be removed from the scene. How you get it out depends on a number of factors.

Is the plane intact and flyable? You could fly it out if the property owner doesn't mind. If you landed on a road, you're probably out of luck. Many areas have ordinances prohibiting takeoffs from public property, including roads.

Are your skills up to it? It's one thing to do soft-field practice on an asphalt runway. It's another thing entirely to try to drag your 172 out of a fallow field with trees in the distance.

Finally, your insurance probably prohibits it. Take a look at your policy. It probably allows operation only from approved airfields.

If you have an in-flight hull policy, it should include a clause covering "transportation to the nearest airport in the event of an off-airport landing." Secure the airplane and contact your insurance company. If you don't have hull coverage, you might have a chance. Your liability coverage might apply. Check your policy or talk to an agent.

Similar factors hold if the aircraft isn't flyable. If the plane is totaled, the insurance company will take possession. If not, your hull coverage should handle transportation. The means of transportation of your aircraft depends upon its proximity to roads. Hauling it from an isolated mountain meadow will probably require a helicopter. Sometimes the plane can be carried intact; otherwise it'll have to be partially disassembled. Leaving wreckage in place might not be an option, either. All the more reason to carry hull coverage.

If the site is accessible by road, recovery will cost quite a bit less. Again, if your insurance covers it, leave it to them. Otherwise, you'll have to either hire a specialist or bring it home yourself. A large trailer is typically pulled to the landing site (Fig. 13-10). The fuel tanks are drained (assuming they weren't empty already), and the wings removed. The fuselage is pulled onto the trailer using either manpower or a winch. If the horizontal stabilizer is wider than highway laws allow, it must be removed, too. External locks are applied to the rudder and elevator, if necessary.

Fig. 13-10. This aircraft recovery trailer was pressed into service for a parade appearance.

The wings are either stowed on the trailer or carried separately. They must be secured during transport, or they might blow off the trailer. Any wooden supports will have to be built back at the airport or prepared at the landing site. Take a portable generator to run a circular saw. Plenty of padding will be necessary if you want to protect the paint job.

You can disassemble the plane and place it on the trailer, but an A&P will have to supervise and sign off the reassembly (Fig. 13-11).

Fig. 13-11. Cessna wings are easily removable for transport, but reinstallation must be supervised and signed off by an A&P.

14

The answer

WHEN YOU BOUGHT this book, you probably had one overriding question: How much does it cost to own a plane?

Ask owners, and some will grin and say, "Everything you have, plus ten percent."

If you press for actual numbers, they might not be able to *give* you an accurate estimate. I talked to a lot of owners while researching this book. Some weren't sure how much their metal steeds drained their pocketbooks. "I don't *want* to know," one said. A case-study participant gave me several pages detailing his expenses over a number of years. At the end was the admonition, "Don't tell my wife!"

Another common thread appeared during the interviews. "I spent a lot on parts until I found this place with great prices . . .." "I really got taken during my first annual, but a guy told me about this other outfit . . .." Like flying, airplane ownership has a learning curve. The longer you own an airplane, the more connections you make, and the cheaper it becomes. So what should you do? Simple: Buy an airplane.

Start looking for a good used 150, 172, or Cherokee. Get a good prepurchase inspection. Keep some money in reserve in case of postpurchase problems.

Fly it for a year, then take a look at your finances. Are you affording it? Is your cash reserve gone? Is the plane just too much of a drag on the budget? If negative answers dominate, sell it. You'll probably *make* money. If you've only flown 100 hours or so and haven't hurt the airplane, you'll have little trouble finding a buyer. Prices are skyrocketing. I sold my 150 for $5,500; 10 years later, the airplanes are going for twice that for good, used 1965 models.

Be warned: There is a danger. When you have owned an airplane, you'll never be able to rent again. Sure you can call the local FBO and schedule yourself some 172 time. But after your 1.3-hour flight ("Return the aircraft 10 minutes early for the next renter."), you'll slowly carry the keys back to the office. You'll listen to the "ka-chunk" of the credit-card machine while the sky is still singing in your blood.

You'll think about those old biplanes flocking around the little strip that the FBO won't let anyone land at. You'll remember watching the family that was picnicking under the wing of a 172. Don't forget the "icky-yet-pleasant" labor of scrubbing the bugs off the leading edge, then watching the gleam of your very own airplane emerge from beneath the suds and bug parts.

You'll be ruined, all right.

As I am. At press time I learned that our club homebuilt was sold. By the time that you read this, Fly Baby N500F will be in the hands of its new owner. It's a moment of concurrent sadness and anticipation.

Sadness in the departure of an old friend that carried me for 8 years and more than 250 flight hours without a significant problem. It always brought me home; it didn't cling leechlike to my bank account. *Au revoir*, Fly Baby.

Anticipation is obvious. Somewhere out there is an airplane with my name on it. I am not sure about the route that I am going to take. My new plane might emerge from my garage after a few years' work. Or scanning the classified advertisements might pay off with the perfect airplane at the right price. Either way, I will be flying again.

The adventure begins anew!

Appendix A

Type clubs

THE FOLLOWING LISTS the addresses of clubs for a variety of aircraft types. When looking for an aircraft of interest, remember that some aircraft were manufactured by a variety of companies; for instance, the Navion was built by both Ryan and North American.

As a rule, type clubs generally issue their own newsletters that range from several photocopied sheets a quarter to monthly magazines. Dues that typically include a subscription range from $5 to $40 a year on up. Circulation ranges from 50 to the thousands. Most offer sample copies to prospective subscribers.

The type clubs listed here were active in the first quarter of 1994 based upon responses to my survey.

Aeronca
(See also **Bellanca** and **Champion**)

International Aeronca Association
"Aeronca Lover's Club"
Buzz Wagner
Box 3
401 1st St. E
Clark, SD 51522

Aeronca Aviator's Club
55 Oakley Avenue
Lawrenceburg, IN 47025

Arctic Tern

Arctic Newsletter
5630 South Washington Road
Lansing, MI 48911-4999

Aviat

Husky Newsletter
5630 South Washington Road
Lansing, MI 48911-4999

Beechcraft

Twin Beech 18 Association
Enrico Bottieri
P.O. Box 8186
Fountain Valley, CA 92708

American Bonanza Society
Mid-Continent Airport
P.O. Box 12888
Wichita, KS 67277

Musketeer Newsletter
5630 South Washington Road
Lansing, MI 48911-4999

Skipper Newsletter
5630 South Washington Road
Lansing, MI 48911-4999

Tomahawk Newsletter
5630 South Washington Road
Lansing, MI 48911-4999

World Beechcraft Society
1436 Muirlands Dr.
La Jolla, CA 92037

Bellanca

Bellanca/Champion Club
Capitol Airport
P.O. Box 708
Brookfield, WI 53008

West Coast Viking Owners Group
Gary Robinson
2640 Obelisco Pl.
Carlsbad, CA 92009

Cessna

Cardinal Club
Phil Harrison
1701 St. Andrew's Drive
Lawrence, KS 66047

Cessna 150/152 Club
P.O. Box 15388
Durham, NC 27704

Cessna 195 Group
Charles A. Spurrell
28608 Coveridge Dr.
Rancho Palos Verdes, CA 90274

Cessna Airmaster Club
9 S 135 Aero Dr.
Naperville, IL 60564-9417

Cessna Owner Organization
P.O. Box 337
Iola, WI 54945

Cessna Pilots Association
P.O. Box 12948
Wichita, KS 67277

Eastern 190/195 Association
25575 Butternut Ridge Rd.
North Olmsted, OH 44070

International Bird Dog Association
3939-C8 San Pedro NE
Albuquerque, NM 87110

International Cessna 120/140
  Association
Box 830092
Richardson, TX 75083

International Cessna 170
  Association,   Inc.
P.O. Box 1667
Lebanon, MO 65536

International 180/185 Club
H. Buz Landry
Box 222
Georgetown, TX 78627

International 195 Club
P.O. Box 737
Merced, CA 95341

The Twin Cessna Flyer
8531 Wealthwood Drive
New Haven, IN 46774

West Coast Cessna 120/140 Club
9087 Madrone Way
Redding, CA 96002

DeHavilland

Beaver/Otter Newsletter
5630 South Washington Road
Lansing, MI 48911-4999

Ercoupe

Ercoupe Owners Club
P.O. Box 15388
Durham, NC 27704

Funk

Funk Association
Gene Ventress
10215 S. Monticello
Lenexa, KS 66227

Globe

(Also **Temco**, et al)

Southern California Swift Wing
19821 Kinzie St.
Chatsworth, CA 91311

Grumman

(Also **American**, **American General**)

AA-1 Newsletter
5630 South Washington Road
Lansing, MI 48911-4999

AA-5 Newsletter
5630 South Washington Road
Lansing, MI 48911-4999

American Yankee Association
(Includes all AA-series, not just Yankee)
P.O. Box 1531
Cameron Park, CA 95682

Lake
(Also **Colonial**)

Lake Amphibian Flyers Club
815 N. Lake Reedy Blvd
Frostproof, FL 33843

Luscombe

Luscombe Association
6438 W. Millbrook
Remus, MI 49340

Continental Luscombe Association
705 Riggs
Emmett, ID 83617

Maule

Maule Aircraft Association
5630 S. Washington Road
Lansing, MI 48910

Mooney

Western Association of Mooney Mites
Anthony A. Terrigno
18020 South Trail
Chino Hills, CA 91709

Navion
(Ryan, North American)

American Navion Society
Box 1810
Lodi, CA 95241-1810

Piper

Arrow Newsletter
5630 South Washington Road
Lansing, MI 48911-4999

Cherokee Pilot's Association
P.O. Box 7927
Tampa, FL 33673

Cub Club
6438 W. Millbrook
Remus, MI 49340

Flying Apache Association
John Lumley
6778 Skyline Dr.
Delrey Beach, FL 33446

International Comanche Society
P.O. Box 400
Grant, NE 69140

International Liaison
   Pilot and Aircraft Association
Bill Stratton
16518 Ledgestone
San Antonio, TX 78232

Piper Owner Society
P.O. Box 337
Iola, WI 54945

Short Wing Piper Club
220 Main St.
Halstead, KS 67056

Super Cub Pilots Association
P.O. Box 9823
Yakima, WA 98909

Rallye

Rallye Newsletter
5630 South Washington Road
Lansing, MI 48911-4999

Ryan
(See also **Navion**)

National Ryan Club
(for Pre-WWII/WWII Ryans)
Bill Hodges
19 Stoneybrook Ln
Searcy, AZ 72143

Stearman

Stearman Restorers Inc.
P.O. Box 10663
Rockville, MD 20849-0663

Stinson

National Stinson Club
115 Heinley Rd.
Lake Placid, FL 33852

Southwest Stinson Club
2264 LosRobles Rd.
Meadow Vita, CA 95722

Taylorcraft

Taylorcraft Owner's Club
12809 Greenbower NE
Alliance, OH 44601

Travel Air

Travel Air Restorers Association
Jerry Impellezzeri
4925 Wilma Way
San Jose, CA 95124

Varga/Shinn/Morrisey

Varga/Shinn VG-21 Squadron
Pat Beery
7845 Soda Bay Rd.
Kelseyville, CA 95451

Varga Newsletter
5630 South Washington Road
Lansing, MI 48911-4999

Waco

American Waco Club
3546 Newhouse Place
Greenwood, IN 46143

Western Waco Association
P.O. Box 706
Groveland, CA 95321

Appendix B

A handpropping guide

YOU'VE PROBABLY SEEN it happen. A small homebuilt or an older classic sits ready to fly, someone steps up to the propeller (Fig. B-1), and somewhere in the gathered crowd a voice says: "I don't believe she's doing that." She gives the prop an easy fling, and the engine starts. The plane taxis away, leaving a sea of shaking heads. "No way . . .. There is no way you'd catch me handpropping."

Fig. B-1. Many deem handpropping dangerous, but the risks are acceptable with proper training. (Never attempt to handprop an airplane without having received proper training from experienced personnel.)

Most of them are worried about the dangers. And I'll admit, turning an airplane engine by hand is more dangerous than mashing a button labeled START. Yet starters have been around so long, we've forgotten what they cost. Additionally, the typical starter weighs about 20 pounds, and you'll need a battery, so add 20 more pounds. Complexity increases as well: wires, circuit breakers, and switches.

That complexity adds to your maintenance costs. Think back to the problems you've had with rental aircraft. How many problems have been electrical in nature? It's pretty simple: Buy a plane without an electrical system, and you won't have any electrical problems.

I know, the thought of touching that 6-foot prop just gives you the willies, doesn't it? A lot of people feel that way. So many, in fact, that it's getting tougher and tougher to sell airplanes that don't have starters.

There is the opportunity. If the demand is less, the price drops. If you're willing to learn to handprop, you can buy good airplanes for thousands of dollars less. Handpropping is not that tough. With proper precautions and procedures, the typical lightplane engine can be easily and safely started by hand.

We'll cover these precautions and procedures, but understand that you cannot learn handpropping from a book. Study up, then find an A&P or other pro to get some dual instruction. The process is pretty easy, but personalized instruction is imperative.

SOME MECHANICAL BASICS

Three things are necessary to start an aircraft engine: fuel, air, and spark. Rotation, as such, isn't needed. Load the proper fuel-air mix into a cylinder, position the prop to the right spot, and fire a spark. The engine will kick over and start. Some older engines actually use this system.

Magnetos are used on aircraft because they're self-contained; they need no outside source of power to generate the spark. But like an alternator or generator, the magnetos must be turning to work. With the mags geared to the engine crankshaft, the prop must rotate for the magnetos to spark. It doesn't matter if this rotation is provided by Bendix, Delco, or Harry Handpropper.

But Bendix and Delco can't say "Ouch!" To keep this word—and stronger turns of phrase—from one's vocabulary, a deeper understanding of aircraft magnetos is necessary.

Proper spark timing is vital to any internal combustion engine. Magnetos fire the plug at about 25–30° before the piston reaches the top of the compression stroke, which is called *top dead center* (TDC). At cruise RPM, this allows the fuel-air explosion to build prior to the power stroke. But at starting speeds, 30° before TDC is too early because the force of the explosion can kick the prop backwards.

Magneto impulse couplings solve this problem. At speeds less than 400 RPM or so, the impulse coupling retards the spark (i.e., makes it later) for easier starting. When engine RPM approaches idling speed, timing shifts back to normal. Impulse magnetos make electric starters feasible for small aircraft; otherwise the starter motor would have to be big enough to turn the prop at near-idle speeds.

When handpropping, it's important to know which, if any, mags have impulse couplings. On most engines, you'll hear a distinct click or snap if the prop is slowly rotated. The sound is produced by the other function of impulse couplings: momentarily snap the magneto forward to produce a hotter spark. If there's only a single click, only one mag (probably the left) has a coupling. When in doubt, ask an A&P.

Impulse couplings make handpropping a breeze. The prop doesn't have to be turned quickly, just fast enough so that inertia carries it through the compression stroke. In fact, flip it too fast, and the coupling disengages. With only one impulse coupling, turn the other mag OFF during start.

Engines without couplings require special care. If the prop isn't turned fast enough, the spark will sometimes kick the propeller backwards. This kickback can actually turn the engine several rotations in reverse. Poor handpropping technique can result in bruised or broken hands and wrists.

Let's look at the right way to do it.

TURNING TECHNIQUE

The following is aimed at garden-variety Continentals and Lycomings. Antique and classic engines, such as Kinners and Rangers, might require special procedures. First, we'll examine prop turning technique, then cover the overall process involved in actually starting the airplane.

Practice with the switch OFF until you get the hang of it. This is where an experienced trainer can make a difference because he can spot what you're doing wrong. Before beginning, ensure the prop is in the correct position. Normally, props will come to a stop just before a compression stroke. This usually places one blade at about the 10 o'clock position as viewed from in front. On a typical taildragger, the blade ends up at about shoulder height or a little higher.

If the prop isn't in the correct position, turn off the mags and gingerly rotate the prop backward. This keeps the impulse coupling from firing and reduces the chances of a cylinder having an explosive fuel-air mixture in case the mag does fire. Remember that the mags are turned off by shorting the primary coils. While flying, that's great because if the switch wire breaks, the engine keeps running. But if it happens on the ground, the mag can fire anytime the prop is moved. So rotate it backward when possible, and never turn it forward without proper handpropping technique.

(Some engine-driven vacuum pumps don't like to be turned backward. If the engine has a pump, *carefully* turn the prop forward.)

Proper technique starts with standing in front of the prop to the left of the hub (your left, not the aircraft's) with one foot slightly forward. Lift your left or right foot, whichever is more comfortable. Stand close enough so that you can comfortably reach the prop by leaning forward slightly. A deadly mistake by neophytes is to stand too far away, resulting in an awkward lean toward the prop. Being a little scared is natural and healthy, but balance is crucial. Lean too far, and a minor slip or bobble might pitch you into a spinning propeller.

Place both hands on the blade about two-thirds the distance out from the hub. This position gives good mechanical advantage without requiring a lot of arm motion. It also places your hands close to the widest part of the blade.

Don't wrap your fingers around the trailing edge (Fig. B-2). That's bad news if the prop kicks back. Instead, move the prop by the friction between your palms

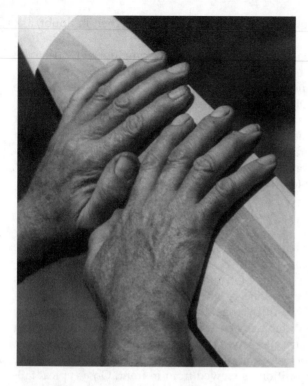

Fig. B-2. Pull the prop through with the assistance of friction between your fingers and the blade, not by wrapping fingertips around the trailing edge.

and the blade. The propeller's angle of incidence and hand pressure downward combine to give a very good grip. Bare hands work best.

To prop the engine, sweep your arms in a short arc downward, curving your hands clear in the follow-through. The trick is to move your arms away when the prop's own momentum is enough to complete the stroke. You aren't trying to spin the engine over several times; you just want to get it moving fast enough to move past one cylinder's compression stroke. Because the propeller also acts as a fly-wheel, you can release the blade quite early. By the time the fuel-air mixture fires, your hands are clear of the prop arc.

As you pull the prop through, rock backward, and get your torso moving away from the prop. There's nothing wrong in turning this motion into a step or two backward. You can even "push off" the blade a little bit to get yourself moving. This is quite natural in a taildragger because the tilted plane of the propeller tends to bring your arms toward you as they're brought downward.

If the engine doesn't have an impulse mag, turn the prop as rapidly as possible while still keeping your balance. If the aircraft has an impulse coupler, just a gentle flip will do.

STARTING PROCEDURE

Handpropping is much more than just turning the propeller. Let's assume that you go through the entire starting process without a qualified person in the cockpit. We'll cover two-person start procedures later.

Too often pilots are so worried about being hit by the prop that they ignore the other major danger of handpropping: the runaway aircraft. Tie it down! Period. Locally in the last few years, a J-3 chewed up an airport office, a Bonanza with a dead battery creamed a Baron, a nose-dragging VariEze center punched a Cherokee, and a Cessna 180 careened between two rows of aircraft for a half hour until the prop finally hit something hard enough to stop it.

At the very minimum, tie the tail and use a pair of chocks. Set the parking brake, if equipped. Even better, have an A&P install a glider hook and cockpit release (Fig. B-3). They're not all that expensive and allow you to release the tail rope from the pilot's seat.

Fig. B-3. This glider hook on the tailwheel allows the aircraft to be secured during handpropping. The hook is released via a cockpit handle.

Before you touch the prop, check your footing. You're going to be standing near a sharp, rapidly-turning object. You don't want to slip. Ice is an obvious hazard, but loose gravel, wet grass, or wet leaves can be just as bad. Clear the ground or roll the plane somewhere else.

With the ground clear and the airplane secure, we're finally ready to start. If the engine is cold, a few primer shots will be necessary. All engines are different, but our club Fly Baby's C-85 likes one shot if the air temperature is between 55–70°F, two shots between 45–55°F, and three shots below 45°F. Don't prime if the engine has run in the last couple of hours or so.

Make sure the mags are off and the throttle is closed. Walk to the front of the airplane and tug on the prop near the hub to verify the plane is secure. Flip the prop through about six or seven blades to distribute the priming charge and loosen the oil.

Walk back to the cockpit. Set the throttle closed or slightly cracked. Turn the appropriate mag(s) to the ON position. Glance at the chocks and the tiedowns to ensure everything's secure.

Back at the front, tug on the hub again. Take your stance and flip the prop, remembering to follow through smoothly. Unless it starts, the prop will bounce between two compression strokes for a moment. Take care because when the engine is balky, it's too easy to automatically reach forward after each pull. The engine can fire anytime the prop is moving, and you might end up with broken hands or worse. Wait until it stops completely, then step up for another try.

Occasionally the prop will come to a sudden stop because the engine has hung up in the middle of a compression stroke. *This is very dangerous*; the engine could fire with the least motion. Turn off the mags before repositioning the prop.

When the engine finally starts, walk in a wide arc around the front of the airplane, and walk toward the cockpit. Get into the habit of walking around the wing, even if the plane is a high-winger; too many pilots have tried to walk through the prop.

An alternate starting procedure is used for some airplanes, notably Piper Cubs. The J-3's two-piece door folds completely out of the way. It's possible to stand in front of the open door and turn the propeller from the back side. You turn the prop with your right hand, while holding onto the door frame with your left. When the engine starts, all you have to do is lean into the cockpit to get access to the switches and throttle (Fig. B-4).

Whichever way you start the engine, you can perform your normal post-starting tasks when you have access to the controls: turn on the other mag, adjust idle, check oil pressure, and the like. When satisfied, back off the throttle to minimum RPM, pull the chocks, and undo the ropes.

Figure B-5 summarizes the handpropping procedure.

STARTING PROBLEMS

Our Fly Baby's Continental typically starts on the second or third flip. So should your engine, if it's in reasonable shape. But what if it doesn't start? Assuming nothing's broken, starting problems are usually traced to one source: fuel, either too much (flooded) or too little (starved).

It's obvious when the engine is flooded. The engine either doesn't fire or just sputters, and raw gasoline often drips from the air inlet. To clear a flooded engine, turn the mag switches OFF and open the throttle all the way (pushed completely forward). Flip the engine backward for 10 or so blades to clear the excess gas. Then close the throttle, and turn on the mags.

Use care with this procedure. Make sure the throttle is closed (pulled completely back) before attempting to start. The engine can and will start at full throttle. I made this mistake once. *Once.* Fortunately the plane was tied down, but I went into adrenaline overload before reaching the cockpit.

If you are trying to start an engine that is fuel-starved, it starts, runs for a little while, then dies. The solution is more gas, either by cracking the throttle (pushing

Fig. B-4. This homebuilt owner is propping from behind because the position gives him access to the engine controls in the cockpit. This is a popular method for planes with open cockpits or split doors, such as the Piper J-3.

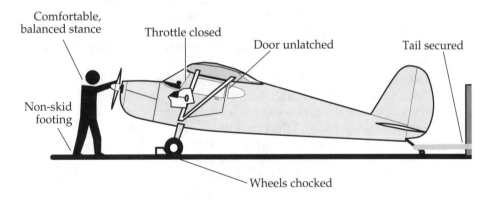

Comfortable, balanced stance

Throttle closed

Door unlatched

Tail secured

Non-skid footing

Wheels chocked

Fig. B-5. Overview of handpropping procedure

it slightly forward) or by additional priming. Too much, though, will flood the engine. Use care because it's easier to add a little more fuel than to clear a flood.

In cold weather, don't blast the engine with many shots of prime. The raw gas strips oil from the cylinder walls and causes high wear upon startup. Instead, shoot another dose and leave the primer knob unlocked. Start the engine, and add little squirts when the engine begins to falter.

If only one mag has an impulse coupling, and the engine refuses to fire, the magneto might be a little weak. Try starting on both mags. Flip the prop rapidly, as if neither has a coupling.

TWO-PERSON START PROCEDURE

Propping an airplane by yourself isn't really the best procedure. Like I mentioned earlier, the major safety problem is the runaway airplane. The ideal solution is to find someone to prop your engine while you sit in the cockpit. But the handproppers of yesterday aren't around anymore. And with the liability situation, you should be a bit leery of accepting a volunteer whose skills might be rusty or nonexistent.

While it's hard to find a qualified person to prop your engine, it isn't too difficult to find a pilot or mechanic willing to sit in the cockpit while you do the work. Nothing says the assistant actually has to do anything because you can tie down the aircraft and set the switches and throttle as if no one were there. Brief your assistant on the brakes and engine controls in case something goes awry.

If you can trust the person to actually work the engine controls, the whole process is much easier. Here's the procedure:

1. Brief the helper on the sequence of activities and the commands to be used.
2. Seat the helper in the cockpit.
3. Prime the engine appropriately.
4. Step to the prop.
5. Call out, "Brakes set, throttle closed (or cracked), switches off."
6. The helper checks the brakes, throttle, and switches, and verbally verifies each: "Brakes set, throttle closed (or cracked), switches off."
7. Tug on the prop hub to make sure the brakes are set, and turn the prop through six or seven blades.
8. Call "Brakes set, throttle closed, contact." **Note:** Use "contact" instead of "switches on" because it is too easy to mishear a "switches on/off" command.
9. The helper should ensure the brakes and throttle are set, call "Brakes set, throttle closed, contact," then turn on the mag switches.
10. Tug on the hub again, then start the engine normally.

If you want to terminate the start procedure, call "Switches off." The helper should turn off the mags, then call, "Switches off."

THE LEGAL SIDE OF SOLO HANDPROPPING

One of the continuing controversies about handpropping is whether you can legally start an airplane without a pilot or A&P in the cockpit. There are no FARs that specifically address the issue; however, FAR 91.13, "Careless or Reckless Operation" is a useful catchall. Indeed, published FAA guidelines specify a license suspension of 30–90 days for "propping an aircraft without a qualified person at the controls."

However, no FAA representative that I spoke to in the safety, enforcement, and legal sections interpreted 91.13 in that fashion. All felt solo starts are safe as long as the aircraft is tied down and chocked. (None of the representatives would go on record.)

The violation guideline is probably intended for those cases where an accident results. If you let the plane run away, consider a violation the least of your worries. Find a pilot to sit in the cockpit while you start the engine. In any case, tie the plane down!

HANDPROPPING SUMMARY

The first time you hand-start an engine is a real thrill; there is something "Waldo-Pepperish" about it. But continued safe operation demands good habits as well as good technique. For instance, after turning on the mags, always pull on the throttle to ensure it's closed. On the way to the prop, "twang" the tiedowns to verify that they are tight and secure. Initiate a poststart ritual of always patting a wingtip with your hand; you'll be less likely to walk into the running prop. Allow no distractions; encourage no conversation.

Like every aircraft procedure, currency is important. Many handprop accidents involve aircraft with dead batteries. If you're well-practiced, go ahead and prop it. Otherwise, it's far safer to dig up a pair of jumper cables. (If you run the battery dead on a Lycoming, expect a struggle. The starter gear remains engaged, and you'll have to fight the drag of the starter while turning the prop.)

Obviously, handpropping is far more involved than turning the key to START. But do you really need an electrical system? Sure, you'd look rather silly flipping the prop of a Piper Arrow. But if a handheld navcom meets your avionics requirements, an airplane without an electrical system can save you a bundle.

There's a small amount of additional risk involved. But it's minimized by careful preparation and appropriate procedures. Considering the money that you can save, and the performance that you can gain, why not hand prop? Contact!

Glossary

100LL One-hundred octane low-lead aviation fuel.

100-hour inspection Similar to an annual inspection; performed by an A&P.

A&P Person with an airframe and powerplant rating; authorized to perform most inspection and maintenance on aircraft.

accessories Officialese for engine-driven items not built by engine manufacturer: magnetos, starters, alternators, and the like.

accessory case Portion of the engine that contains gearing to turn aircraft accessories, such as magnetos and alternator.

ACO Aircraft certification office.

AD (airworthiness directive) An FAA announcement mandating certain inspections or repairs of given aircraft or equipment.

Airknocker Nickname for the Aeronca Champ.

AN Air Force-Navy standard for aviation hardware.

annual inspection FAA-required in-depth inspection and analysis of aircraft condition. The inspection must be performed by an A&P with an IA authorization.

antique An aircraft built prior to World War II.

AOPA Aircraft Owners and Pilots Association.

autogas Automobile gasoline.

autogas STC Aircraft has an FAA approval to run on automobile gas.

avionics Aircraft electronics, typically for communications and navigation.

basket case Slang term for a disassembled aircraft that needs restoration from the ground up.

chocks Devices to prevent aircraft wheels from rolling when parked.

cigarette Slang term for ceramic insulator at end of aircraft spark plug wire.

classic An aircraft built from 1945–1955. A *contemporary classic* was built from 1956–1962.

communal hangar Large building storing multiple aircraft with a single door for aircraft entry and exit. Planes often have to be repositioned to allow others ingress or egress.

conventional gear Landing gear configuration where the main wheels are mounted well forward, and a smaller wheel is underneath the rudder.

CS or C/S Constant-speed propeller.

DAR Designated airworthiness representative.

DER Designated engineering representative.

direct costs Costs associated directly with flight: fuel and oil expenses, and the like.

EAA Experimental Aircraft Association.

ELT Emergency locator transmitter.

endless loop See recursion.

ethanol A form of alcohol that cannot be present in automobile gasoline that is used in an aircraft that has an autogas STC.

experimental A certification category covering a number of subcategories. The term usually refers to amateur-built (homebuilt) aircraft.

FAA Federal Aviation Administration.

FAR Federal Aviation Regulations.

FAR Part 43 Section of FARs governing performance of aircraft inspections and maintenance.

FAR Part 91 Section of FARs governing aircraft operations and frequency of inspections and maintenance.

FBO Fixed-base operator.

FCC Federal Communications Commission.

ferry permit FAA authorization to fly an aircraft that does not currently meet airworthiness requirements to a location where it can be repaired.

fixed costs Those costs that are connected with the ownership, and not operation, of the aircraft: hangar rent or tiedown fee, insurance premiums, annual inspections, and the like.

flat-rate annual Annual inspection performed for a given price. Price might actually be higher if necessary repairs exceed a certain level.

FSDO FAA flight standards district office.

GPS (global positioning system) Satellite-based navigation system.

groundloop Rapid ground maneuver performed by out-of-control taildraggers that occasionally damages the airplane.

handprop Starting an aircraft engine by manually turning the propeller.

homebuilt Aircraft built by a private individual for education or recreation.

hull coverage Insurance coverage for actual physical damage to the aircraft.

IA A&P with an inspection authorization who is allowed to approve aircraft modifications and major repairs, in addition to performing annual inspections.

impulse coupling Portion of magneto that alters ignition timing to aid starting.

kitplane Homebuilt airplane that is constructed from a prepared kit rather than from raw materials.

leaseback Allowing your aircraft to be rented out in exchange for a given hourly payment.

liability coverage Insurance that covers damage to passengers or third-parties in the event of an accident or incident.

life-limited Aircraft part that must be replaced at a prescribed interval, for example, hydraulic hoses that deteriorate over time.

loran Navigation system designed for maritime use became popular for aviation as well. Loran is losing popularity to GPS.

MDH Major damage history.

milk stool A description of tricycle-gear aircraft.

mogas Automobile gasoline.

MTBE Methyl tertiary butyl ether, which is an automobile gas additive that is acceptable in fuel used by aircraft approved to use autogas.

multimeter Small electrical tool used to measure voltage or electrical resistance.

navcom Radio with navigation and two-way communications capability.

not-in-flight Insurance that covers damage to the aircraft as long as the damage occurred while the aircraft wasn't flying: vandalism, weather damage, and the like.

NTSB National Transportation Safety Board.

OEM Original equipment manufacture: OEM identifies replacement parts made by the same company that provided the original parts.

oil screen Coarse metal mesh that filters engine oil.

overhaul To restore an engine to a condition where it should last its listed TBO.

PMA Parts manufacturer approval: PMA describes replacement parts approved by the FAA to replace parts originally manufactured by another company.

prop strike An incident where the propeller struck something while the engine was turning. (See sudden stoppage.)

ragwing A description of a fabric-covered airplane.

recursion See endless loop.

repairman certificate Authorization to perform a given set of maintenance tasks. A person who constructed a homebuilt aircraft receives a repairman certificate that allows that person to perform all maintenance and inspections on the aircraft.

Rotax A brand of two- and four-stroke engines that are popular on ultralights and small homebuilts.

runout An engine that has reached its TBO.

SDR Service difficulty report. SDRs are the details of unusual maintenance findings that are filed with the FAA by mechanics. The reports are regularly disseminated to the aviation community and might form the basis for a service bulletin or an AD.

service bulletin An airframe and/or engine manufacturer's announcement that certain repairs or inspections should be performed to forestall certain failures. It is similar to an AD, but a service bulletin does not have the force of law.

service manual Official publication by an airframe or engine manufacturer that specifies maintenance procedures.

SBS (side-by-side) Aircraft seating arrangement where pilot and passenger sit next to each other, rubbing elbows.

signature loan A no-collateral loan based solely upon the applicant's credit rating.

SMOH Engine time since major overhaul

STC Supplemental type certificate. An amendment to the original TC, often by a source other than the original manufacturer.

sudden stoppage Term for when the propeller strikes something that causes the engine to halt. The stoppage induces major internal stresses, and the engine usually has to be torn down and inspected.

taildragger An aircraft with "conventional" gear, which is two large wheels mounted forward and a smaller wheel underneath the rudder.

tandem Aircraft seating arrangement where the pilot and passenger sit one behind the other.

TBO Time between overhauls.

TC Type certificate.

TDC Top dead center: when a piston is up as far as it can go inside a cylinder.

tiedown Unsheltered place where aircraft can be parked for a long period of time and secured against wind.

title search Process by which the legal owner of an aircraft is determined. Should be performed prior to purchase, to ensure the seller is the legal owner.

top overhaul An overhaul performed only to the compression section on an aircraft engine (pistons, rings, cylinders, cylinder heads, valves, etc.).

tricycle gear Landing gear configuration where the aircraft's third wheel is mounted in front, like a child's tricycle.

trigear See tricycle gear.

TSO Technical standard order: Basic standard that aircraft parts much meet.

type certificate Document officially defining an aircraft's equipment and configuration. Aircraft must conform to be legally airworthy.

Wanttaja An aviation writer, pronounced "Wahn-TIE-ah."

Index

Other Bestsellers of Related Interest

The Illustrated Buyer's Guide to Used Airplanes, 3rd Edition
—Bill Clarke
The only way for most private pilots to own an airplane is to buy a used one. Clarke points out that there are a lot of makes and models available on the used aircraft market, and deciding which one to buy can be a risky endeavor. In this popular guide, he provides the practical, realistic advice used aircraft buyers need to make smart choices. He covers all aspects of buying and owning a used airplane and includes tips on aircraft insurance, storage, maintenance, and protection.
0-07-011266-5 $21.95 Paper

The Pilot's Guide to Affordable Classics, 2nd Edition
—Bill Clarke
Bill Clarke offers solutions that enable budget-conscious pilots to fulfill their dreams. He shows used airplane buyers that they CAN purchase an aircraft at an affordable price. Provided are pre-purchase inspection tips and information on which aircraft to choose and how to find the desired model. Plane reconditioning, maintenance suggestions, avionics descriptions, and explanations of insurance and paperwork are all part of this invaluable reference.
0-07-011268-1 , $18.95 Paper

How to Make Your Airplane Last Forever
—Mary Woodhouse and Scott Gifford
This manual provides plenty of innovative ideas for increasing airplane longevity by using the most recent techniques available in an environment in which technology and regulations are ever-changing. Individual chapters cover FAA regulations regarding maintenance and repairs, cleaning and storage, troubleshooting aircraft systems, operating tips for extending life, do-it-yourself maintenance, the annual 100-hour inspection, and more.
0-07-071704-4 $21.95 Paper

Cutting the Cost of Flying
—Geza Szurovy
This volume gives more than 200 practical, easy-to-follow suggestions that can save literally thousands of dollars a year for the prudent owner or renter pilot. Included are actions pilots can take to optimize flying time, budget rental strategies, low-cost sources for pilot supplies, finance strategies, and scores of additional ways to legally and safely beat the ballooning costs of flying.
0-07-062993-5 **$16.95 Paper**

The Joy of Flying, 3rd Edition
—Robert Mark
This enthusiastic introduction to the world of piloting invites the reader to experience the beauty, freedom, and joy of flying. Find out how to choose a flight instructor who's right for you. Learn what type of aircraft you want to fly. Get the facts in medical and FAA requirements, radio communications, pilot ratings, partnerships and clubs, aviation organizations, and more.
0-07-040487-9 **$15.95 Paper**

The Plane & Pilot International Aircraft Directory
—Editors of Plane & Pilot Magazine
This outstanding international directory or light aircraft is an easy-to-use, one-stop reference for anyone interested in airplanes typically flown by general aviation pilots. For each aircraft, this fact-filled book lists airframe and powerplant information, historical data, and standard data and performance figures such as: operating speeds, seating and fuel capacities, engine make and model, and performance specifications. Photographs of each plane are included.
0-07-050304-4 **$34.95 Hard**
0-07-050305-2 **$24.95 Paper**

Standard Aircraft Handbook, 5th Edition
—Larry Reithmaier, Ed.
Since 1952, the Standard Aircraft Handbook has been the definitive reference for aviation mechanics engaged in building, maintaining, overhauling, or repairing metal aircraft. Now in its fifth edition, this classic handbook has been updated to include the latest in aircraft parts, equipment, and construction techniques.
0-07-157642-8 **$12.95 Paper**

The Right Seat Handbook: A White-Knuckle Guide to Light Planes
—Charles F. Spence
The Right Seat Handbook gives non-pilots practical information they can use to leave their fears on the ground and fly in small planes with confidence. Spence explains how an airplane flies and what a pilot does in the cockpit, describing such things as how to: read VFR and IFR charts, use the plane's computer to estimate arrival times, and locate the nearest airport. A helpful glossary is included.
0-07-060148-8 **$11.95 Paper**

How to Order

 Call 1-800-822-8158
24 hours a day,
7 days a week
in U.S. and Canada

 Mail this coupon to:
McGraw-Hill, Inc.
P.O. Box 182067
Columbus, OH 43218-2607

 Fax your order to:
614-759-3644

 EMAIL
70007.1531@COMPUSERVE.COM
COMPUSERVE: GO MH

Shipping and Handling Charges

Order Amount	Within U.S.	Outside U.S.
Less than $15	$3.50	$5.50
$15.00 - $24.99	$4.00	$6.00
$25.00 - $49.99	$5.00	$7.00
$50.00 - $74.49	$6.00	$8.00
$75.00 - and up	$7.00	$9.00

EASY ORDER FORM— SATISFACTION GUARANTEED

Ship to:

Name _____

Address _____

City/State/Zip _____

Daytime Telephone No. _____

Thank you for your order!

ITEM NO.	QUANTITY	AMT.

Method of Payment:

☐ Check or money order enclosed (payable to McGraw-Hill)

☐ DISCOVER

☐ AMERICAN EXPRESS Cards

☐ VISA

☐ MasterCard

Shipping & Handling charge from chart below	
Subtotal	
Please add applicable state & local sales tax	
TOTAL	

Account No. ☐☐☐☐☐☐☐☐☐☐☐☐☐☐☐☐

Signature _____ Exp. Date _____
Order invalid without signature

In a hurry? Call 1-800-822-8158 anytime, day or night, or visit your local bookstore.

Key = BC95ZZA

629.133   Wanttaja, Ron.
W
        Airplane ownership.

$24.95   09/18/1996

DATE			

BAKER & TAYLOR